T0183234

Springerbriefs in Molecular Science

Chemistry of Foods

Series Editor
Salvatore Parisi, Lourdes Matha Institute of Hotel Management and Catering Technology, Thiruvananthapuram, Kerala, India

The series Springer Briefs in Molecular Science: Chemistry of Foods presents compact topical volumes in the area of food chemistry. The series has a clear focus on the chemistry and chemical aspects of foods, topics such as the physics or biology of foods are not part of its scope. The Briefs volumes in the series aim at presenting chemical background information or an introduction and clear-cut overview on the chemistry related to specific topics in this area. Typical topics thus include:

- Compound classes in foods—their chemistry and properties with respect to the foods (e.g. sugars, proteins, fats, minerals, …)
- Contaminants and additives in foods—their chemistry and chemical transformations
- Chemical analysis and monitoring of foods
- Chemical transformations in foods, evolution and alterations of chemicals in foods, interactions between food and its packaging materials, chemical aspects of the food production processes
- Chemistry and the food industry—from safety protocols to modern food production

The treated subjects will particularly appeal to professionals and researchers concerned with food chemistry. Many volume topics address professionals and current problems in the food industry, but will also be interesting for readers generally concerned with the chemistry of foods. With the unique format and character of SpringerBriefs (50 to 125 pages), the volumes are compact and easily digestible. Briefs allow authors to present their ideas and readers to absorb them with minimal time investment. Briefs will be published as part of Springer's eBook collection, with millions of users worldwide. In addition, Briefs will be available for individual print and electronic purchase. Briefs are characterized by fast, global electronic dissemination, standard publishing contracts, easy-to-use manuscript preparation and formatting guidelines, and expedited production schedules.

Both solicited and unsolicited manuscripts focusing on food chemistry are considered for publication in this series. Submitted manuscripts will be reviewed and decided by the series editor, Prof. Dr. Salvatore Parisi.

To submit a proposal or request further information, please contact Dr. Sofia Costa, Publishing Editor, via sofia.costa@springer.com or Prof. Dr. Salvatore Parisi, Book Series Editor, via drparisi@inwind.it or drsalparisi5@gmail.com

More information about this subseries at https://link.springer.com/bookseries/11853

Ramesh Kumar Sharma · Maria Anna Coniglio ·
Pasqualina Laganà

Natural Inflammatory Molecules in Fruits and Vegetables

 Springer

Ramesh Kumar Sharma
Food Safety Consultant and Scientific
Writer
Tilam Sangh Rajasthan
Bikaner, Rajasthan, India

Maria Anna Coniglio
Department of Medical, Surgical Sciences
and Advanced Technologies "GF Ingrassia"
University of Catania
Catania, Italy

Pasqualina Laganà
Department of Biomedical and Dental
Sciences and Morphofunctional Imaging
University of Messina
Messina, Italy

ISSN 2191-5407   ISSN 2191-5415   (electronic)
SpringerBriefs in Molecular Science
ISSN 2199-689X   ISSN 2199-7209   (electronic)
Chemistry of Foods
ISBN 978-3-030-88472-7   ISBN 978-3-030-88473-4   (eBook)
https://doi.org/10.1007/978-3-030-88473-4

This Springer imprint is published by the registered company Springer Nature Switzerland AG
The registered company address is: Gewerbestrasse 11, 6330 Cham, Switzerland

Contents

Chapter 1
Fruits and Vegetables, Though Rich in Antioxidants, Might Lead to Cytotoxicity

Abstract The antioxidant and anti-inflammatory properties of fruits and vegetables make them distinctly placed and renowned food articles. They are rich in vitamins, minerals, polyphenols, and fibres, also having high water contents. It is now believed that despite being rich in antioxidants, fruits and vegetables might lead to cytotoxicity because of improper ways of farming, storage, and processing. The cytotoxicity appears to be related to the depletion of the human body's glutathione level as a result of consumption of poor diet—both the lack of fruits and vegetables and the inclusion of improperly farmed or stored or processed fruit-vegetable commodities—leading to infection, chronic diseases, and constant inflammation. Glutathione is considered as one of the body's important antioxidants because it is able to considerably reduce oxidative stress via the decomposition of free radicals in the body. This molecule is a tripeptide composed of three amino acids—cysteine, glutamic acid, and glycine—present in mammalian tissues, which acts as a free-radical scavenger and detoxifying agent. Not only the improperly farmed or stored or processed fruits and vegetables, but also vegetables containing appreciably high amounts of naturally occurring glycoalkaloids like potatoes, tomatoes, eggplants, and peppers might lead to depletion of the body's glutathione level and give rise to inflammation. This chapter is an attempt to present a scenario of fruits-vegetables which despite being rich in antioxidants can cause cytotoxicity, based on current researches.

Keywords Alkaloid · Bacterial wilt · Cytotoxicity · Glutathione · Natural inflammatory molecule · Reactive nitrogen species · Reactive oxygen species

Abbreviations

DPPH	2,2-Diphenyl-1-picrylhydrazyl
AP	Active principle
CaC_2	Calcium carbide
DNA	Deoxyribonucleic acid
GSH	Glutathione
HDL	High-density cholesterol

© The Author(s), under exclusive license to Springer Nature Switzerland AG 2022   1
R. K. Sharma et al., *Natural Inflammatory Molecules in Fruits and Vegetables*, Chemistry of Foods, https://doi.org/10.1007/978-3-030-88473-4_1

·OH   Hydroxyl free radical
MDA   Malondialdehyde
mM   Millimol
NO   Nitric oxide
NPK   Nitrogen, phosphorus, and potassium
OC   Organochlorine
OP   Organophosphorus
GSSG   Oxidised glutathione
RNS   Reactive nitrogen species
ROS   Reactive oxygen species
·O–O·   Superoxide free radical
–SH   Thiol

1.1 Introduction

It is a common belief that growth and energy providing protein and lipid molecules, excessively taken by the body via food consumption, are good targets for oxidative attack, whose modification in the body can increase the risk of inflammation giving rise to many diseases including cancer, neurological disorders, atherosclerosis, hypertension, diabetes, respiratory problems, and asthma [1]. In normal conditions, it appears that oxidative interactions with fat and protein molecules resulting in the formation of free radicals such as reactive oxygen species are almost necessary to fight against microbial infections and protect the body from bacterial and viral infections. It is because the new agents to reach clinical practice are reactive oxygen species (ROS), which are effective against most Gram-positive and Gram-negative organisms [2].

The oxidative stress is the result of an imbalance between the production of oxidants—reactive metabolites or free radicals, normally called ROS and reactive nitrogen species (RNS)—and their elimination by protective mechanisms referred to as antioxidation or reduction systems. Oxidant agents, ROS and RNS, include superoxide free radicals (·O–O·), hydroxyl free radicals (·OH), and nitric oxide (NO), which are of course balanced by enzymatic antioxidants like superoxide dismutase, catalase, and glutathione peroxidase; the inflammation or oxidative stress condition dominates in the body when there is an imbalance in favour of the oxidants [3, 4]. The optimum consumption of fruits and vegetables is normally considered helpful in the prevention of inflammation-causing oxidative stress or maintenance of oxidation–reduction balance in the body. It appears that the fruits and vegetables, if consumed in excess, perhaps might lead to an imbalance in favour of the antioxidants in the body, making it prone to infections.

The consumption of unhealthy foods is considered as one of the reasons for chronic inflammation in the body. Foods able to cause inflammation are bakery items made of refined carbohydrates, fried items containing rancid oil, soda and other sugar-sweetened beverages, red meat (burgers, steaks), processed meat (hot dogs, sausage), margarine, shortening, and lard [5]. It means that avoidance of refined cereal flours or consuming whole grain flours, avoidance of rancid oil or use of fresh edible oils and fats in preparations in the kitchen, and avoidance of sugar-sweetened beverages, red meat, processed meat, margarine, shortening, and lard are perhaps a good food use policy. However, the unrefined carbohydrates and fresh oils and pulses too can create a mild surplus oxidation condition; food items of antioxidant or anti-inflammatory nature seem to be incorporated in the diet. Despite some different opinions about priority in listing anti-inflammatory food, nowadays these are listed as follows [6]:

(1)   Fresh fruits, including grapefruit, grapes, blueberries, bananas, apples, mangoes, peaches, tomatoes, and pomegranates.
(2)   Dried fruits, including plums (prunes).
(3)   Vegetables, especially broccoli, Brussels sprouts, cauliflower, and bok choy.
(4)   Vegetable products with notable protein amounts, such as chickpeas, *seitan*, and lentils.
(5)   Fatty fish, such as salmon, sardines, albacore, tuna, herring, lake trout, and mackerel.
(6)   Whole grains, including oatmeal, brown rice, barley, and whole wheat bread.
(7)   Leafy greens, including *kale*, spinach, and romaine lettuce.
(8)   Ginger.
(9)   Nuts, including walnuts and almonds.
(10) Seeds, such as chia and flaxseed.
(11) Food filled with ω-3 fatty acids, such as avocado and olive oil.
(12) Coffee.
(13) Green tea.
(14) Dark chocolate (moderate consumption).
(15) Red wine (moderate consumption).

This list of anti-inflammatory foods suggested by Moria Lawler [6] is based on some reliable sources [2, 7–9]. Except for fatty fish (these foods are ranked fifth as priority), the list does not include any other non-vegetarian food item as an essential anti-inflammatory food. It appears that this listing of anti-inflammatory foods is based on a particular study, which reports that reducing inflammation by following a vegan or vegetarian diet may help delay disease progression, reduce joint damage, and potentially diminish reliance on rheumatoid arthritis medication when used as a complementary therapy [10]. Fruits and vegetables occupy the first three priorities in the aforementioned list of anti-inflammatory foods. The quintessence is that antioxidant and anti-inflammatory properties of fruits and vegetables make them distinctly placed and renowned food articles.

Diets with high fruits and vegetable contents are widely recommended for their health-promoting properties [11]. Fruits and vegetables have historically held a place in dietary guidance because of their concentrations of vitamins, especially vitamins C and A; minerals, especially electrolytes; and more recently phytochemicals, especially antioxidants (polyphenols) [11]. In addition, fruits and vegetables are a source of dietary fibres [11]. Furthermore, fruits and vegetables are a good source of water in addition to potable water, keeping the body well hydrated; limited dehydration can cause fatigue, headaches, skin problems, muscle cramps, low blood pressure, and a rapid heart rate while prolonged dehydration can lead to serious complications like organ failures [12]. Fresh fruits and vegetables normally contain around 90% water.

It is now believed that fruits and vegetables might lead to cytotoxicity because of improper ways of farming, storage, and processing, despite being rich in antioxidants and several nutrients. This chapter briefly deals with these toxicity-causing factors and makes a comprehensive discussion on how cytotoxicity depletes the body's glutathione level in the context of current researches to reach the conclusion and how some of the properly farmed, stored, and processed fruits and vegetables too might exert cytotoxic effects in the human body due to ozone disinfection (carried out to cope with microbial contamination) and presence of natural inflammatory molecules. This discussion also concerns briefly ROS and their importance as inflammatory agents by natural sources.

1.2   Cytotoxic Effects of Improperly Farmed, Stored, and Processed Fruits and Vegetables

Toxicity can refer to the effect of poison or chemical on a whole organism, such as an animal, bacterium, or a plant, as well as the effect on a substructure of the organism, such as the liver (hepatotoxicity); it is found to be dose-dependent and species-specific [13]. Since the basic structural unit of every living thing is the cell, the cytotoxicants (substances toxic to cells) appreciably affect the growth and devel-opment of the organism. Fruits and vegetables are nowadays too improperly farmed, stored, and processed to remain quite safe for humans. Modern intensive (industrial) agricultural methods have stripped increasing amounts of nutrients from the soil in which the food, we eat, grows; each successive generation of fast-growing, pest-resistant fruits and vegetables is truly less good for the consumer than the one before, as it suffers from a decline in the amount of protein, phosphorus, iron, riboflavin (vitamin B_2), and vitamin C [14]. Along with the soil depletion aspect of inten-sive farming, the crisis of pesticides application in gardens and farms is also looming large and affecting public health. Nowadays, pesticides are used on fruits, vegetables, wheat, rice, olives and canola pressed into oil, and on non-food crops such as cotton, grass, and flowers; organophosphorus (OP) pesticides malathion and chlorpyrifos are commonly used on all fruits, vegetables, and wheat [15].

As far as the toxicity aspects of synthetic pesticides are concerned, it is now almost established that major pesticides (mixture of pesticides) are more toxic to human cells than their declared active principle (AP), which is the only one tested in the longest toxicological regulatory tests performed on mammals, mostly rats and mice [16]. The cytotoxic effects of malathion and other commonly used OP pesticides on fruits and vegetables are quite evident; methamidophos, malathion, and malaoxon cause significant inhibition of cell viability and increased genetic damages in cell line PC 12 derived from pheochromocytoma at 40 mg/l [17]. Due to this reason, genotoxicity and carcinogenicity and serious health effects of malathion (O,O-dimethyl-S-1,2-bis ethoxy carbonyl ethyl phosphorodithionate) intake too are now considered along with common symptoms of toxicity including numbness, tingling sensation, headache, dizziness, difficulty breathing, irritation of the skin, exacerbation of asthma, abdominal cramps, and death [18]. Another OP pesticide—chlorpyrifos, commonly applied on fruits and vegetables, and currently banned in Hawaii, New York, and California States—is known for its damaging effects on the human nervous system, similarly for other OP pesticides. It blocks an enzyme called acetylcholinesterase that the brain needs to control acetylcholine, one of the many neurotransmitters mediating communication between nerve cells [19]. Continuous exposure can lead to respiratory paralysis and death [19]. Although the main target organ for chlorpyrifos is the nervous system, this pesticide is also worldwide known for cytotoxicity and endocrine disruption; chlorpyrifos and its principal metabolite 3,5,6-trichloro-2-pyridinol have been evaluated by means of in vitro assays [20].

A prominent feature of intensive industrial farming of fruits is the early cultivation of unripe fruits and their artificial ripening. In spite of health warnings from the government, unripe fruits are subjected to the process of ripening with a chemical ripening agent (calcium carbide) in storages to meet our market demand. Several in vivo studies have reported the toxicological outcomes such as histopathological changes in lungs and kidneys and haematological and immunological responses, upon exposure to calcium carbide (C_aC_2) [21]. An in vitro study indicates a 23.14% reduction in cell viability at 0.2 g/l; even a short-term C_aC_2 exposure may increase the cellular oxidative stress and disturb the redox balance of the cell which then undergoes apoptosis [21]. Nowadays, fruits are also processed for their convenient usage. Powdered juices are widely consumed by the population because of their convenient preparation; however, they have complex formulations, consisting of several classes of food additives. In the 2010s, a study evaluated the toxicity at the cellular level of industrialised powdered juices of orange and guava flavours, concluding that both juice powders promoted significant anti-proliferative effects on *Allium cepa* L. root meristem cells and exhibited cytotoxic and genotoxic potential [22].

1.3   Cytotoxicity Due to Fruits and Vegetables Pesticides Can Deplete Body's Glutathione Level

Organophosphate pesticides are generally regarded as safe for use on crops (pest control) and animals (contagion control) due to their relatively fast degradation rates [23]. The other type of frequently applied pesticide is organochlorines, which are known for their high toxicity, slow degradation, and bioaccumulation [24]. Malathion and chlorpyrifos, generally applied on fruits and vegetables, belong to organophosphorus insecticides and undoubtedly are less toxic and fast biodegradable compared to organochlorines (OC). However, these OP pesticides are potent cholinesterase inhibitors that reversibly or irreversibly bind covalently with the serine residue in the active site of acetyl cholinesterase and prevent its natural function in the catabolism of neurotransmitters [25]. Nowadays, their cytotoxic effects are known, and researchers link them with glutathione depletion cytotoxicity.

A recent experimental study clearly observes that malathion increases levels of a lipid peroxidation product, malondialdehyde (MDA), and reduces glutathione (GSH, an antioxidant, detoxifying agent in mammalian tissues) content; however, it is reported to reduce MDA level and increase GSH content in rat ovarian tissues if used in combination with ascorbic acid [26]. This study concludes that malathion induces lipid peroxidation and oxidative stress, reducing glutathione levels in the ovarian of rats; in addition, it appears that ascorbic acid can recover malathion-induced poisonous changes due to its antioxidant nature [26]. The authors of this book would like to present a hypothesis: fruits like lemons and oranges that contain appreciable amounts of vitamin C (ascorbic acid) are safer than sweet fruits like papaya, watermelon, or muskmelon if subjected to the same levels of intensive farming, at least for the reproductive system.

In 2003, Indian researchers Radhey Shyam Verma and co-workers studied the effect of chlorpyrifos on thiobarbituric acid reactive substances, scavenging enzymes, and glutathione in rat tissues [27]. According to this study, chlorpyrifos exposure generates oxidative stress in the body, as evidenced by an increase in thiobarbituric acid reactive substance (lipid peroxidation products) and a decrease in the levels of superoxide scavenging enzymes namely superoxide dismutase, catalase, and glutathione peroxidase in liver, kidney, and spleen at all doses; levels of GSH-reduced form decrease, while levels of oxidised glutathione (GSSG) increase [27]. This simple discussion leads to the fact that cytotoxicity due to fruits and vegetable pesticides can deplete glutathione levels in the human body.

1.4 Glutathione Molecule. Configuration and Physiological Activity

Despite being naturally enriched with various nutrients, including antioxidants like vitamin C and various polyphenols, valuable agricultural products such as fruits and vegetables nowadays might contain appreciable amounts of inflammatory synthetic molecules belonging to artificial pesticides like malathion and chlorpyrifos, ripening agents like calcium carbide, and some processing aids (particularly in fruit juices). These synthetic inflammatory molecules present in fruits and vegetables are presumed to affect body glutathione levels in the human body. Glutathione molecule present in the plants, animals, fungi, and some bacteria cells acts as an important antioxidant or free-radical scavenger and detoxifying agent [28]. In the human body, cell mechanism maintains intracellular redox balance. Its chemical configuration is as follows: a tripeptide composed of three amino acids namely cystein, glutamic

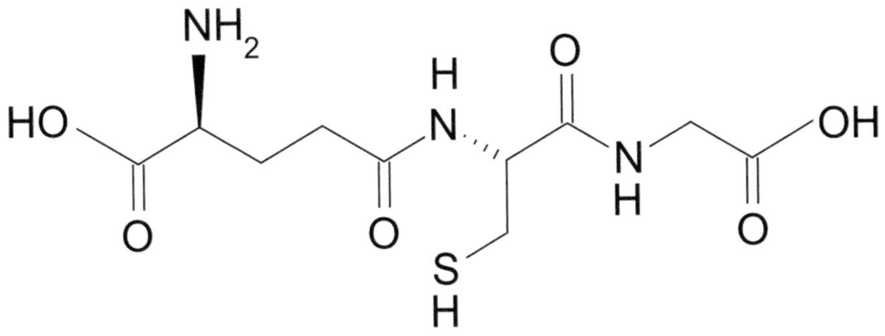

Fig. 1.1 The molecular structure of glutathione (γ-*L*-glutamyl-*L*-cysteinylglycine)

acid, and glycine. Glutamic acid gets attached via its side chain to the N-terminus of cysteinylglycine [28, 29], as shown in Fig. 1.1.

Glutathione, present both in the cytosol and the organelles, is the most abundant thiol molecule in animal cells, ranging from 0.5 to 10 mM, taking into account that 1 millimol (mM) of glutathione is equal to 0.3073 g [29]. The sulphur-hydrogen bond in the thiol (–SH) group of glutathione molecule plays the important role in the redox reaction. GSH molecule protects cells by means of chemical reduction of present ROS [29]. For example, the reduction of peroxides by GSH might be illustrated in the form of redox reaction as follows:

$$2\,GSH + R_2O_2 \leftrightarrows GSSG + 2\,ROH \tag{1.1}$$

where R is for a general alkyl group, GSH is the reduced form, and GSSG is the oxidised form.

The official nomenclature of GSH indicating sulfanyl bond is as follows: (2s)-2-amino-5 (((2R)-I-(carboxymethylamino)-1-oxo-3 sulfanylpropan-2yl) amino)-5-oxopentanoic acid [30]. The reduced form of glutathione—the antioxidant form present in micromolar (μM) concentration in bodily fluids and in mM concentrations in tissues—participates in the detoxification of hydrogen peroxide by means of various glutathione peroxidases [31]. The GSH/GSSG ratio is an indicator of cellular health, with reduced form (GSH) constituting up to 98% of cellular GSH under normal conditions [31]. However, GSH/GSSG ratio can be correlated with the occurrence of neurodegenerative diseases, such as Parkinson's and Alzheimer's disease [31].

1.5 Fruits and Vegetables, Though Rich in Antioxidants, Might Lead to Cytotoxicity

Consumption of fruits and vegetables, properly grown in a natural way, owing to the presence of minerals, vitamins, and antioxidant polyphenols, might lead to elevation of antioxidant level in the human body, potentially helping the person prevent the inflammation or cell damage associated with oxidative stress. The interesting point concerned with the antioxidant action of fruits and vegetables is to know a brief detail of natural antioxidants contained by those, which perhaps fight against peroxides or oxidants produced by protein and lipid oxidation and are helpful in GSH conservation in the body.

It is now accepted that blueberries possess a class of plant compounds called anthocyanins, which give blueberries both their blue colour and many of their health benefits; blueberries can help in heart health, bone strength, skin health, blood pressure, diabetes management, cancer prevention, and mental health because one cup of this fruit provides 24% of the recommended daily allowance of vitamin C per capita [32]. The artichokes—a Mediterranean thistle-like vegetable plant, known as *chukandar, vajrangi,* or *hathichak* in India—are often seen on menus in dips or on the top of salads and are regarded as a superfood in the sense they are naturally rich source of vitamins A, K, C, B$_6$, thiamine, riboflavin, niacin, folate, calcium, iron, zinc, potassium, magnesium, and phosphorus [33].

Research has shown that artichoke can help and strengthen the immune system, lower cholesterol, and detoxify the liver, and it may also protect against cancer, diabetes, heart attacks, and strokes; their high-fibre content can help ease digestive issues, reduce blood pressure, and even eliminate hangovers [33]. The pecan nut, known as *bhidurkashtha phal* (fruit) in *hindi,* belonging originally to the Southern United States and Mexico, is regarded as an antioxidant-packed free-radical fighter, which may lower cholesterol [34]. A single cup of chopped pecan nuts contains 132 mg of bone-strengthing magnesium, 111 mg of phytosterols, and enough amounts of vitamin E, fibres, and mono- and polyunsaturated fat which can help reduce the risk of heart disease and type-2 diabetes [34]. Packed with vitamins, fibre, manganese, potassium, and particularly high levels of antioxidants known as polyphenols, strawberries are a sodium-free, fat-free, cholesterol-free, and low-calorie food; these potent little packages protect the heart, increase high-density cholesterol (HDL) with interesting cardiovascular benefits, lower blood pressure, and guard against cancer [35]. It has been found that daily consumption of strawberries results in a modest but significant antioxidant capacity in a healthy population [36].

It is said about red cabbage that its organic compounds are too many to list, but essential components include B-vitamins, calcium, manganese, magnesium, iron, potassium, vitamin C, A, E, K, dietary fibre, and antioxidants like anthocyanins and indoles; it is known for anti-cancer, anti-ageing, weight loss, eye care, bone density, anti-Alzheimer's disease, and immune system boosting potentials [37]. A single cup of raspberries provides over 50% of the minimum daily target for vitamin C, which supports immunity and skin health and helps produce collagen; they also contain manganese and vitamin K, which both play a role in bone health [38]. Resveratrol—one of the various antioxidant polyphenols, present in grapes in combination with vitamin C, β-carotene, quercetin, lutein, lycopene, and ellagic acid—remarkably lowers blood pressure and protects against the development of cancer [39]. The green leafy vegetable spinach, being high in insoluble fibre, may benefit digestion; it also packs high amounts of carotenoids, vitamin C, vitamin K, folic acid, iron, and calcium, required, respectively, for eyesight, skin health and immune function, blood clotting, women's pregnancy state, haemoglobin formation, nervous system, and cardiac health [40]. Research findings show that carotenoids found in dark green leafy vegetables such as *kale* can act as antioxidants and boost the body's own antioxidant defences and help stop free radicals from damaging deoxyribonucleic acid (DNA) that can lead to cancer [41].

The antioxidant effects of dietary carotenoids present in carrots and other vegetables may reduce the risk of cancer; carrots appreciably contain vitamin A which helps a person to see in the dark [42]. It is a common belief that several other popular fruits like apples, oranges, bananas, peaches, pears, mangoes, papaya, watermelon, muskmelon, blackberry (Java plum), plums, etc. (commonly included in the human diet) are rich in antioxidants, vitamins, and minerals. On the other side, the intensive industrial agricultural practices followed by gardeners or farmers nowadays have almost broken this belief. Non-preserved soils, belonging to a region where biodiversified dense forests are absent, are too loosely packed to withstand heavy rains; they undergo a process—called soil erosion—in which a fraction moves along with rainwater and ultimately goes to the sea turning seawaters gradually rich in minerals and obviously making the mineral profile of soil and groundwater poorer and poorer [43].

Along with the aspect of weakening soil mineral profile, the other aspect of intensive agriculture is the application of synthetic pesticides and fertilisers in farms or gardens. Nowadays, some countries, particularly India, are tending towards minimisation of synthetic pesticide application [44]; as a result, the probability of their presence in such countries is very low. Since India and almost all countries where organic farming practices are followed are passing from climate change or global warming, and the probability of the presence of aflatoxins (fungal mycotoxin) in crops is obviously somewhat higher [43]. The local warming factor or contribution to the global warming of India and several organic farming-practising countries appear to be much dominant due to vast deforestation. This situation obviously gives rise to increasing soil erosion (when heavy rainfall takes place and subsequent mineral nutrients deficiency takes place [43].

Of course, macronutrient minerals belonging to elements such as nitrogen, phosphorus, and potassium (NPK) are available to plants grown via NPK synthetic fertilisers along with the supplementation of iron, copper, zinc, and chromium. A few micronutrient minerals are also nowadays prevalent; however, the supplementation of the entire 72-element mineral range required for human body maintenance is perhaps expensive or non-efficient in the agriculture of countries like India [43].

As a result, crops of countries like India suffer from serious mineral deficiency and immune system weakness with chances of bacterial, fungal, and viral pathogenic developments in crops, particularly with relation to fruits and vegetables containing high water content [43].

Although the list of contamination typologies in food articles is remarkable, just a few types are developed at the farm level due to industrial agricultural practices (pesticides), abnormal weather conditions (mycotoxins and pathogens), non-preserved soil and water situations (poor micronutrient mineral profile), and toxicity of particular natural constituents in crop or food (alkaloids) [43]. Heavy metals, although optimally required for conservation by farm soil and groundwater used for irrigation, may be excessively ejected out in natural water streams by industries; the World Health Organization enlists arsenic, cadmium, lead, and mercury among the top ten chemicals or group of chemicals of major public health concern [43].

As a result of food frauds, hormones like oxytocin have been often observed recently in milk, vegetable, and fruits [43]. Perhaps, these are the major contamination factors due to which fruits and vegetables, despite being rich in antioxidants, might lead to cytotoxicity. The toxicity of particular natural constituents called alkaloids in crop or food say soybean, coca, opium poppy, coffee, tea, kola nuts, cocoa beans, and tobacco and fruits-vegetables like potatoes, tomatoes, and eggplants too is worth discussing.

1.6 Current Researches on Cytotoxicity of Commonly Consumed Antioxidant-Rich Fruits and Vegetables

Green leafy vegetables are perhaps most commonly consumed as the part and parcel of a healthy diet in the world because of their good antioxidant property, appreciable presence of a variety of minerals and vitamins, and feeble occurrence of natural inflammatory molecules like alkaloids. However, in 2009, Sri Lankan origin researchers Gayathri Balasuriya and H. Ranjith W. Dharmaratne working in Kolbe University and University of Mississippi, respectively, conducted cytotoxicity tests for green leafy vegetables using brine shrimp (*Artemina salina*) lethality bioassay. As anticipated, the majority of tested leafy vegetables were found to have insignificant cytotoxicity; however, some of *Aerva lanata* and *Baccopa monnieri* greens showed a significantly higher level of cytotoxicity when compared with the positive control. *Alternanthera sessilis* (which is the most popular leafy vegetable among Sri Lankans) and *Passiflora edulis* showed similar toxicity levels as the positive control. According to researchers, consumption of these four leafy vegetables could pose a potential health risk, and further toxicological studies should be carried out to evaluate their potential health risks [45].

It is worth noting that the antioxidant activity of the above-mentioned four greens were tested using 2,2-diphenyl-1-picrylhydrazyl (DPPH) assay, and all the tested leafy vegetables showed free-radical scavenging properties indicating the presence of primary antioxidants in plants [45].

In the light of the above-mentioned current research scenario on the cytotoxic effects of antioxidant-rich green vegetables, the authors of this book infer that several of the commonly consumed fruits and vegetables might yield lower returns of their antioxidant activity than previously established values perhaps due to intensive industrial farming, climate change, deforestation, and improper ways of processing.

1.7 Current Researches on Microbial Contamination in Fresh Fruits and Vegetables

In the context of the safe consumption of fresh fruits and vegetables, it is somewhat said in India that cut fresh fruits and vegetables should be consumed within ten seconds; however, this time limit should depend upon environmental factors, say relative humidity and temperature. On the other side, the current and larger production of fruits and vegetables within the shortest possible time to meet the growing demand has placed them at a higher risk of contamination with pathogens, making the safety of consumers uncertain [46]. Washing and rinsing under running water, the safe technique employed for the removal of pathogenic microbes from the surfaces of fresh fruits and vegetables, is capable to remove a few pathogens, while a good number of them persist even after this treatment [46]. Therefore, ozone disinfection for fresh fruits and vegetables is suggested: for example, 2 ppm of ozonized water treatment to leaf lettuce for about 2 min is found to be effective in the disinfection of the lettuce [47]. In the context of ozone disinfection treatment of fresh fruits and vegetables, the authors of this book intend to infer that excessive ozonisation might produce peroxides and other inflammatory oxidant molecules even in fresh fruits and vegetables.

1.8 Current Researches on Disease in Fruits and Vegetables Containing Natural Inflammatory Molecules

Bacterial wilt—one of the main diseases of nightshades, so-called solanaceous crops, caused by the bacterium *Ralstonia solanacearum*—ends up with substantial losses in crops like tomato, eggplant, potato, banana, etc. [48, 49]. The pathogen *R. solanacearum* is a diverse and complex species whose five races affect different plant species, and six biovars indicate its capability to oxidise hexoses, alcohols, sorbitol, and disaccharides, and four phytotypes, region-wise I Asia, II America, III Africa, and IV Indonesia, are known [50, 51]. The solanaceous crops family—which includes potato, tomato, brinjal, chilli, and capsicum—plays an important role in the human diet and the economy of nations [52]. Unfortunately, pesticide applications are nowadays advised to farmers to cope with the bacterial wilt disease; pesticides include poisonous chemicals as follows: benomyl, carbendazim, flubendazole, propiconazole, chloropicrin, 3-(3-indolyl) butanoic acid, validamycin A, methyl bromide, 1, 3 dichloropentene, etc. [52]. Needless to say, these inflammatory pesticide molecules are additionally applied to nightshade vegetables which naturally contain inflammatory molecules termed as alkaloids, say α-solanine in potatoes, α-tomatine in tomatoes, and solasonine in eggplants. Therefore, the authors of this book would like to present an overview of natural inflammatory molecules in food articles in the context of consumption limitations with a special focus on nightshades family, grown throughout the world but too much suffering from pesticide applications.

1.9 Conclusions

The current research scenario on the cytotoxic effects of antioxidant-rich green vegetables inspires the authors of this book to infer that several of commonly consumed fruits and vegetables might yield lower returns of their antioxidant activity than previously established values perhaps due to intensive industrial farming, climate change, deforestation, improper ways of processing, and excessive disinfection (ozonisation). The nightshades—potatoes, tomatoes, eggplants, chilli, capsicum—which contain natural inflammatory molecules are nowadays subjected to excessive poisonous pesticide applications to cope with bacterial wilt disease and prevent economic losses. The authors intend to present an overview of natural inflammatory molecules present in food articles in the context of consumption limitations with a special focus on nightshades family, grown throughout the world.

References

1. Birben E, Sahiner UM, Sackesen C, Erzurum S, Kalayci O (2012) Oxidative stress and antioxidant defense. J World Allergy Org 5(1):9–19. https://doi.org/10.1097/WOX.0b013e318243 9613
2. Dryden M (2017) Reactive oxygen therapy; a novel therapy in soft tissue infection. Curr Opin Infect Dis 30(2):143–149. https://doi.org/10.1097/QCO.0000000000000350
3. Deponte M (2013) Glutathione catalysis and the reaction mechanism of glutathione-dependent enzymes. Biochim Biophys Acta 1830(5):3217–3266. https://doi.org/10.1016/j.bagen.2012.09.018
4. Pizzino G, Irrera N, Cucinotta M, Pallio G, Mannino F, Arcoraci V, Squadrito F, Altavilla D, Bitto A (2017) Oxidative stress—harms and benefits for human health. Oxid Med Cell Longev 2017:8416763. https://doi.org/10.1155/2017/8416763
5. Hu F (2014) Foods that fight inflammation. Harvard Medical School, Harvard Health Publishing, Harvard. Updated 29 Aug 2020. https://www.health.harvard.edu/staying-healthy/foods-that-fight-inflammation. Accessed 23 June 2021

6. Lawler M (2020) A comprehensive guide to an anti-inflammatory diet. www.eve rydayhealth.com. https://www.everydayhealth.com/diet-nutrition/diet/anti-inflammatory-diet-benefits-food-list-tips/. Accessed 23 June 2021

7. Bustamante MF, Agustín-Perez M, Cedola F, Coras R, Narasimhan R, Golshan S, Guma M (2020) Design of an anti-inflammatory diet (ITIS diet) for patients with rheumatoid arthritis. Contemp Clin Trials Comm 17:100524. https://doi.org/10.1016/j.conctc.2020.100524

8. Obón-Santacana M, Romaguera D, Gracia-Lavedan E, Molinuevo A, Molina-Montes E, Shivappa N, Hebert JR, Tardón A, Castaño-Vinyals G, Moratalla F, Guinó E, Marcos-Gragera R, Azpiri M, Gil L, Olmedo-Requena R, Lozano-Lorca M, Alguacil J, Fernández-Villa T, Martín V, Molina AJ, Ederra M, Moreno-Iribas C, Perez B, Aragonés N, Castello A, Huerta JM, Dierssen-Sotos T, Gómez-Acebo I, Molina-Barceló A, Pollán M, Kogevinas M, Moreno V, Amiano P (2019) Dietary inflammatory index, dietary non-enzymatic antioxidant capacity, and colorectal and breast cancer risk (MCC-Spain Study). Nutrients 11(6):1406. https://doi.org/10.3390/nu11061406

9. Mayo Clinic Staff (2021) The truth behind the most popular diet trends of the moment. Mayo Clinic, Rochester. https://www.mayoclinic.org/healthy-lifestyle/weight-loss/in-depth/the-truth-behind-the-most-popular-diet-trends-of-the-moment/art-20390062. Accessed 23 June 2021

10. Khanna S, Jaiswal KS, Gupta B (2017) Managing rheumatoid arthritis with dietary interventions. Front Nutr 4:52. https://doi.org/10.3389/fnut.2017.00052

11. Slavin JL, Lloyd B (2012) Health benefits of fruits and vegetables. Adv Nutr 3(4):506–516. https://doi.org/10.3945/an.112.002154

12. Elliott B (2017) 19 water-rich foods that help you stay hydrated. Updated 9 Aug 2017. www.healthline.com. https://www.healthline.com/nutrition/zero-calorie-foods#TOC_TITLE_HDR_30. Accessed 23 June 2021

13. Dixit V (2019) A simple model to solve complex drug toxicity problem. Toxicol Res 8(2):157–171. https://doi.org/10.1039/C8TX00261D

14. Anonymous (2011) Dirt poor: have fruits and vegetables become less nutritious? www.scientificamerican.com. https://www.scientificamerican.com/article/soil-depletion-and-nutrition-loss/. Accessed 23 June 2021

15. The Center for Ecogenetics and Environmental Health (2013) Fast facts about health risks of pesticides in food. The Center for Ecogenetics and Environmental Health, University of Washington, Washington, D.C. https://depts.washington.edu/ceeh/downloads/FF_Pesticides.pdf. Accessed 23 June 2021

16. Mesnage R, Defarge N, Spiroux de Vendômois J, Séralini GE (2014) Major pesticides are more toxic to human cells than their declared active principles. Biomed Res Int 2014:179691. https://doi.org/10.1155/2014/179691

17. Lu XT, Ma Y, Wang C (2012) Cytotoxicity and DNA damage of five organophosphorus pesticides mediated by oxidative stress in PC 12 cells and protection by vitamin E. J Environ Sci Health Part B 47(5):445–454. https://doi.org/10.1080/036012342012.663312

18. Tchounwou PB, Patlolla AK, Yedjou CG, Moore PD (2015) Environmental exposure and health effects associated with Malathion toxicity. In: Larramendy ML, Soloneski S (eds) Toxicity and hazard of agrochemicals. IntechOpen, London. https://doi.org/10.5772/60911

19. Anonymous (2021) Chlorpyrifos. Pesticide Action Network North America, Berkeley. https://www.panna.org/resources/chlorpyrifos-facts. Accessed 24 July 2021

20. Echeverri-Jaramillo G, Jaramillo-Colorado B, Sabater-Marco C, Castillo-López MÁ (2021) Cytotoxic and estrogenic activity of chlorpyrifos and its metabolite 3, 5, 6-trichloro-2-pyridinol. Study of marine yeasts as potential toxicity indicators. Ecotoxicology 30(1):104–117. https://doi.org/10.1007/s10646-020-02315-z

21. De I, Kour A, Wani H, Sharma P, Panda JJ, Singh M (2020) Exposure of calcium carbide induces apoptosis in mammalian fibroblast L929 cells. Toxicol Mech Methods 31(3):1–11. https://doi.org/10.1080/15376516.2020.1849484

22. de Sousa BM, de Sousa MG, de Castro E, Sousa JM, Peron AP (2016) Cytotoxic and genotoxic potential of powdered juices. Food Sci Technol 36(1):49–55. https://doi.org/10.1590/1678-457x.0006

23. Sidhu GK, Singh S, Kumar V, Dhanjal DS, Datta S, Singh J (2019) Toxicity, monitoring and biodegradation of organophosphate pesticides: a review. Crit Rev Environ Sci Technol 49(13):1135–1187. https://doi.org/10.1080/10643389.2019.1565554

24. Jayaraj R, Megha P, Sreedev P (2016) Review article. Organochlorine pesticides, their toxic effects on living organisms and their fate in the environment. Interdiscip Toxicol 9(3–4):90–100. https://doi.org/10.1515/intox-2016-0012

25. Barr DB, Buckley B (2011) In vivo biomarkers and biomonitoring in reproductive and developmental toxicity. In: Gupta RC (ed) Reproductive and developmental toxicology. Elsevier, Inc., Amsterdam, pp 253–265. https://doi.org/10.1016/B978-0-12-382032-7.10020-7

26. Arab SA, Nikravesh MR, Jalali M, Fazel A (2018) Evaluation of oxidative stress indices after exposure to malathion and protective effects of ascorbic acid in ovarian tissue of adult female rats. Electron Phys 10(5):6789–6795. https://doi.org/10.19082/6789

27. Verma RS, Srivastava N (2003) Effect of chlorpyrifos on thiobarbituric acid reactive substances, scavenging enzymes and glutathione in rat tissues. Indian J Biochem Biophys 40(6):423–428

28. Anonymous (2021) Glutathione—the amazing detoxification molecule you might not know. www.askthescientists.com. https://askthescientists.com/qa/glutathione/. Accessed 24 June 2021

29. Wu G, Fang YZ, Yang S, Lupton JR, Turner ND (2004) Recent advances in nutritional sciences-glutathione metabolism and its implications for health. J Nutr 134(3):489–492

30. National Center for Biotechnology Information (2021) PubChem compound summary for CID 124886, glutathione. National Center for Biotechnology Information, Bethesda. https://pubchem.ncbi.nlm.nih.gov/compound/Glutathione. Accessed 24 June 2021

31. Owen JB, Butterfield DA (2010) Measurement of oxidized/reduced glutathione ratio. In: Totowa NJ (ed) Protein misfolding and cellular stress in disease and aging. In: Bross P, Gregersen N (eds) Protein misfolding and cellular stress in disease and aging. Methods Mol Biol Methods Protoc 648:269–277. https://doi.org/10.1007/978-1-60761-756-3_18

32. Ware M (2017) Everything you need to know about blueberries. Medica News Today, 05 Sept 2017. https://www.medicalnewstoday.com/articles/287710#nutrition. Accessed 24 June 2021

33. Stoller R (2017) Amazing antioxidants in antichokes. National Foundation for Cancer Research, Rockville. https://www.nfcr.org/blog/blogfoodiefridaysamazing-antioxidants-artichokes/. Accessed 24 June 2021

34. Schaufelberger K (2007) Pecans: antioxidant-packed nuts. WebMD LLC, New York City. https://www.webmd.com/food-recipes/features/pecans-antioxidant-packed-nuts. Accessed 24 June 2021

35. Gabrick A (2008) Nutritional benefits of the strawberry. Diet & Weight Management, Feature Stories. WebMD Archives. WebMD LLC, New York City. https://www.webmd.com/diet/features/nutritional-benefits-of-the-strawberry. Accessed 24 June 2021

36. Henning SM, Seeram NP, Zhang Y, Li L, Gao K, Lee RP, Wang DC, Zerlin A, Karp H, Thames G, Kotlerman J, Li Z, Heber D (2010) Strawberry consumption is associated with increased antioxidant capacity in serum. J Med Food 13(1):116–122. https://doi.org/10.1089/jmf.2009.0048

37. Staughton J (2020) 9 impressive benefits of red cabbage. Organic Fats, Organic Information Services Pvt Ltd. Facts. https://www.organicfacts.net/health-benefits/vegetable/red-cabbage.html. Accessed 24 June 2021

38. Sass C (2019) 7 reasons raspberries are so good for you. Health. https://www.health.com/food/raspberries-nutrition. Accessed 24 June 2021

39. Groves M (2018) Top 12 health benefits of eating grapes. Healthline (Nutrition). https://www.healthline.com/health/food-nutrition/are-grapes-good-for-you#memory-boost. Accessed 24 June 2021

40. Gunnars K (2019) Spinach 101: nutrition facts and health benefits. Healthline (Nutrition). https://www.healthline.com/nutrition/foods/spinach. Accessed 24 June 2021

41. American Institute for Cancer Research (2021) Kale: rich in antioxidants. American Institute for Cancer Research, Washington, D.C. https://www.aicr.org/cancer-prevention/food-facts/dark-green-leafy-vegetables/. Accessed 24 June 2021

42. Ware M (2019) Everything you need to know about blueberries. What are the health benefits of carrots. https://www.medicalnewstoday.com/articles/270191#diet. Accessed 24 June 2021
43. Sharma RK, Parisi S (2017) Toxins and contaminants in Indian foods products. Springer International Publishing, Cham. https://doi.org/10.1007/978-3-319-48049-7
44. De A, Bose R, Kumar R, Mazumdar S (2014) Worldwide pesticides use. In: Targeted delivery of pesticides using biodegradable polymeric nanoparticles. Springer India, New Delhi. https://doi.org/10.1007/978-81-322-1689-6
45. Balasuriya G, Dharmaratre HRW (2009) Cytotoxicity and antioxidant activity studies of green leafy vegetables consumed in Sri Lanka. J Natl Sci Found Sri Lanka 35(4):255–258. https://doi.org/10.4038/jnsfsr.v35i4,1315
46. Balali GI, Yar DD, Dela VGA, Adjei-Kusi P (2020) Microbial contamination, an increasing threat to the consumption of fresh fruits and vegetables in today's world. Int J Microbiol 2020(ID 3029295). https://doi.org/10.1155/2020/3029295
47. KumarV MritunjaySK (2015) Fresh farm produce as a source of pathogens: a review. Res J Environ Toxicol 9(2):59–70
48. CABI/EPPO (1999) Distribution maps of plant diseases. Ralstoniasolanacearum. Map nos., 783–785. CAB International, Wallingford
49. Elphinstone JG (2005) Bacterial wilt disease and the Ralstonia solanacearum species complex. In: Allen C, Prior P, Hayward AC (eds) The current bacterial wilt situation: a global overview. American Phytopathological Society, S. Paul, pp 9–28
50. Chandrashekara KN, Prasannakumar MK, Deepa M, Vani A, Khan ANA (2012) Prevalence of races and biotypes of Ralstonia solanacearum in India. J Plant Prot Res 52(1):53–58. https://doi.org/10.2478/v10045-012-0009-4
51. Fegan M, Prior P (2005) How complex is the Ralstonia solanacearum species complex. In: Allen C, Prior P, Hayward AC (eds) The current bacterial wilt situation: a global overview. American Phytopathological Society, S. Paul, pp 449–461
52. Manda RR, Addanki VA, Srivastava S (2020) Bacterial wilt of solanaceous crops. Int J Chem Stud 8(6):1048–1057. https://doi.org/10.22271/chemi.2020.v8.i6o.10903

Chapter 2
Natural Inflammatory Molecules in Fruits and Vegetables: Alkaloids, Uric Acid, and Fructose

Abstract Out of several energy-rich organic compounds formed via photosynthesis in autotrophic organisms like plants, a few of them might prove inflammatory or cytotoxic molecules for humans if consumed regularly. Using previously stored chemical energy, chloroplasts convert atmospheric carbon dioxide mainly into glucose and a little fructose which—in the next step—are combined to yield sucrose, starch, cellulose, etc. Nitrogenous compounds, including proteins, are produced in the consequent step in plants. Purines are nitrogenous molecules that, on regular consumption, can create pathogenesis of gout and hyperuricemia in humans. The cytotoxic effects of foods rich in fructose and purines are well known. On the other hand, the plant defence system produces certain compounds which protect the plants against microorganisms and insects and, on regular consumption, might create inflammation in the human body. The set of plant defensive inflammatory molecules include a variety of alkaloids. In this chapter, an attempt is made to present an overview of natural inflammatory molecules—fructose, purines, and alkaloids—contained in fruits and vegetables.

Keywords Alkaloid · Fructose · Photosynthesis · Plant defence system · Purine · Reactive nitrogen species · Salvage pathway

Abbreviations

CO_2	Carbon dioxide
$C_6H_{12}O_6$	Glucose
O_2	Oxygen
hV	Photon
RNS	Reactive nitrogen species
ROS	Reactive oxygen species
H_2O	Water

2.1   Introduction

Autotrophic nutrition is a process in which the organisms produce their food or photosynthesise complex organic compounds from simple inorganic materials such as water, carbon dioxide, and mineral salts in the presence of sunlight; all green plants are autotrophs or self-feeding organisms [1]. Plants provide food to all herbivores including humans by trapping the solar energy and converting it into chemical energy. Chloroplasts, the part of plant cells that carry out photosynthesis, convert previously stored light energy to chemical energy being stored in root, leaf, seed, or fruit, in the form of mainly glucose and a little fructose which in the next step are combined to yield sucrose, starch, cellulose, etc. [2]. Nitrogenous or nitrogen-containing compounds called amino acids, amides, imino acids, proteins, quaternary ammonium compounds, and polyamines are produced in further next steps, as the result of fixation of atmospheric nitrogen via bacteria like cyanobacteria and utilisation of nitrogenous soil material by plants for their immediate requirements [3]. Purines and uric acid are nitrogenous molecules in plants that can create pathogenesis of gout and hyperuricemia in humans [4]. Cytotoxic inflammatory effects of natural molecules, fructose and purines present in foods, even in fruits and vegetables, are well known. The consumption of fructose—the natural sweetener which is present in fruits—has gained increased attention as a potential cause of hyperuricemia since fructose metabolism produces urate as a by-product, suggesting that patients suffering from gout or hyperuricemia might also avoid or limit intake of fresh fruits, not only foods rich in purines like pulses and beans but also vegetables asparagus, spinach, cauliflower, and mushrooms [5, 6].

On the other hand, certain plant defensive compounds which protect the plants against certain microorganisms and insects, and are by-products of metabolic processes, might create inflammation in the human body if consumed regularly. The set of plant defensive inflammatory molecules include a variety of alkaloids: pyrrolizidine alkaloids (in honey), tropane alkaloids (in wolfberry, cape gooseberry, and coca), isoquinoline alkaloids (in lotus), glycoalkaloids (in potatoes, tomatoes, and eggplants), purine alkaloids (in coffee, tea, kola nuts, and cocoa beans), etc. In this chapter, an attempt is made to present an overview of natural inflammatory molecules—fructose, purines, and alkaloids—contained in fruits and vegetables [7].

2.2   Natural Inflammatory Molecules in Fruits and Vegetables

The autotroph plants capture solar radiation and transform it into chemical or food energy via photosynthesis process, in which atmospheric carbon dioxide is combined with water molecules to yield mainly glucose and a little fructose molecule with the evolution of oxygen molecules; in this way, plants provide oxygen gas required for breathing by all live objects whether animals or vegetation.

In due course of time, glucose molecules get converted into starches. Glucose reacts with fructose to form a sucrose (sugar) molecule, particularly in sugarcane. However, some unreacted fructose remains in fruity plants and mainly exists in fruits providing excellent sweetness and is called fruit sugar. Available studies show that consumption of very sweet fruits might induce purines capable of uric acid generation causing mitochondrial oxidative stress in the human body and triggering etiology of diabetes, hyperuricemia (gout), and obesity [8]. Needless to say that purines are synthesised from simple nitrogenous molecules via de novo pathways and also through degradation of big complex nitrogen-containing molecules via salvage pathways; purines (mostly salvaged) are appreciably, though moderately, found in some vegetables like asparagus cauliflower, spinach, green peas, mushrooms, etc. [9].

The plant immunity or defence system fights against microorganisms and insects and synthesises various complex molecules called alkaloids which possess medicinal values but can cause inflammation if regularly consumed [7].

2.2.1 Photosynthetically Produced Fructose

Chlorophyll—the green colouring pigment of plant leaves—absorbs blue and red parts of sunlight emitting the green radiation and gets excited to fulfil energy requirements for completion of basic photosynthesis reaction, in which atmospheric carbon dioxide (CO_2) and water (H_2O) molecules react to yield glucose ($C_6H_{12}O_6$) and oxygen (O_2) molecules. According to Emerson and Arnold, around 2,500 chlorophyll molecules absorb eight photons (hV) or light quanta, and one molecule of oxygen is released in this photosynthesis process [10].

The balanced photosynthesis equation is as follows:

$$6CO_2 + 6H_2O + 48n(h\nu) \xrightarrow{h\nu} C_6H_{12}O_6 + 6O_2 \tag{2.1}$$

Of course, fructose is also formed along with the formation of glucose in photosynthesis. Naturally occurring glucose is mostly found in ripe grapes, honey, and very sweet fruits, while naturally occurring fructose is found in all fruits and honey. Both the carbohydrate compounds are hexoses with the same molecular formula $C_6H_{12}O_6$. On the other side, they differ in the constitution; the former contains an aldehyde group and is dextrorotatory – D-(+)-isomer called aldohexose or dextrose or D-glucose, while the latter contains a keto group and mostly is levorotatory (ketohexose, laevalose, or L-fructose). Some dextrorotatory fructose, called D-(−)-fructose, is also found in fruits and honey. According to Finar, the only important ketohexose is D-(−)-fructose with melting point 102 °C, the open-chain structure being HO-CH_2.CHOH-CHOH-CHOH-CO-CH_2.OH [11]. The importance of D-(−)-fructose perhaps lies in the fact that the sucrose molecule of cane sugar is composed of D-(+)-glucose and D-(−)-fructose. Free fructose available in fruits is largely levorotatory,

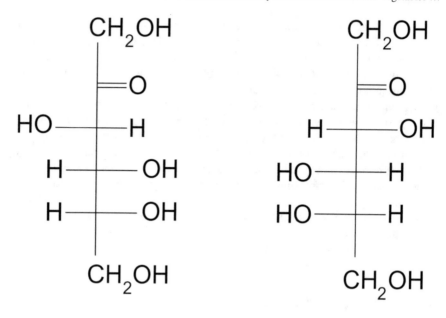

D-fructose   L-fructose

Fig. 2.1 The importance of *D* (−) fructose perhaps lies in the fact that the sucrose molecule of cane sugar is composed of *D* (+)-glucose and *D* (−)-fructose. But free fructose available in fruits is largely laevorotatory and *D*-fructose is found in small amounts [12]. Structurally, fructose is a 2-ketohexose [13]; the accepted open-chain structures of both *D*- and *L*-optical isomers of fructose are shown here

and D-fructose is found in small amounts [11]. From the structural viewpoint, fructose is a 2-ketohexose; the accepted open-chain structures of both D- and L-optical isomers of fructose are shown in Fig. 2.1.

The breakdown of fructose, in the human body, results in the release of purines and ultimately in the formation of uric acid which has a connection with gout and obesity [12].

2.2.2 Salvage Pathways Compounds. Purines

Being participants in many biochemical processes, building blocks for nucleic acid synthesis, and precursors for the synthesis of energy sources like di- and poly-saccharides, purine and pyrimidine nucleotides are of fundamental importance in

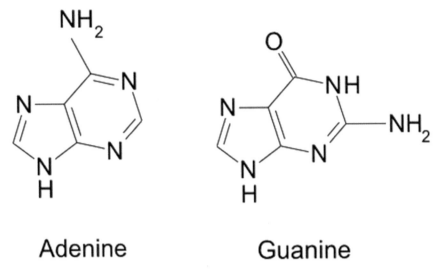

Fig. 2.2 Adenine and guanine. These two nitrogenous bases are of major concern in the diets with the viewpoint of gout connection, with xanthine and hypoxanthine [16]

the growth and development of plants [13]. Purine and pyrimidine nucleotides are synthesised both from amino acids and other small molecules via de novo pathways and from performed nucleobases and nucleosides by salvage pathways [13]. Adenine, guanine, xanthine, and hypoxanthine are four nitrogenous bases of major concern in the diets with the viewpoint of gout connection. Chemical configurations of these four purines are shown in Figs. 2.2 and 2.3. Some non-vegetarian diets such as (a) fish, seafood, and shellfish including anchovies, sardines, mackerel, scallops, herring, mussels, codfish, trout, and haddock; (b) meats such as bacon, turkey, veal, venison, liver, beef kidney, brain, and sweetbreads; and (c) alcoholic beverages correspond to high-purine contents. Also, some vegetables such as asparagus, spinach, green peas, mushrooms, and cauliflower contain appreciable—of course, moderate—amounts of purines [9]. The metabolism of purines in the human body releases uric acid correlated with gout and obesity. The structure of uric acid released by the metabolism of purines is displayed in Fig. 2.4.

2.2.3 Defensively Produced Alkaloids

Numerous complex molecules, belonging to some particular groups, are synthesised in plants as defence attempts against attacks of microorganisms and insects; obviously, they possess medicinal values but can cause inflammation if the concerned foods are regularly consumed [7]. These alkaloid groups are commonly found in borage, comfrey, honey, and also in other foods.

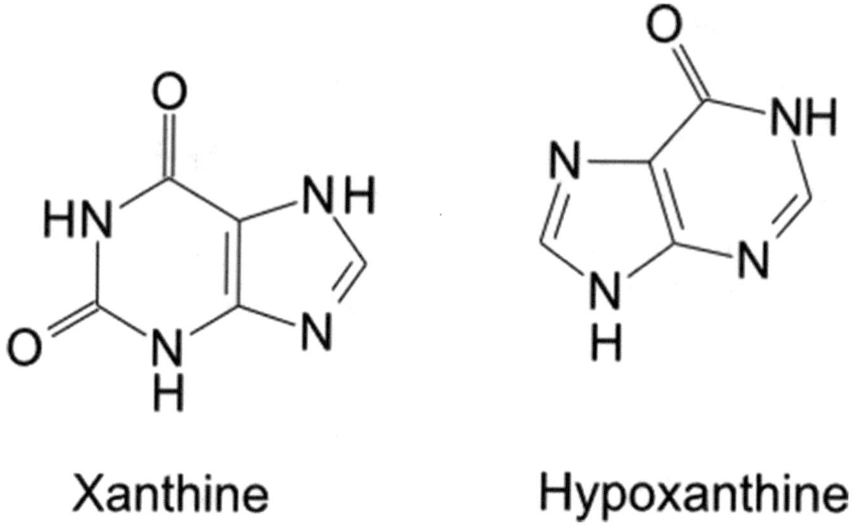

Xanthine   Hypoxanthine

Fig. 2.3 Xanthine and hypoxanthine. These purinic molecules are of major concern in the diets with the viewpoint of gout connection, with adenine and guanine [16]

Fig. 2.4 The structure of uric acid released by the metabolism of purines

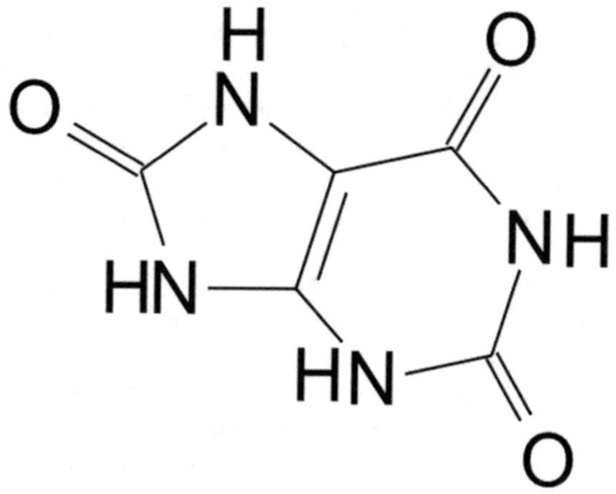

2.3 Natural Inflammatory Molecules Contents in Fruits and Vegetables and Safety Parameters

Despite being enriched with vitamins and minerals and polyphenol antioxidants, fruits and vegetables contain variable amounts of natural inflammatory compounds, namely fructose, purines, and alkaloids. Vitamins, minerals, and polyphenol antioxidants make fruits and vegetables quite safe for human consumption; the so-called natural inflammatory molecules exhibit potent bioactivities and desired effects but are also known for some undesired or possible toxic effects, particularly in the bodily condition of elevated oxidative stress. These molecules might tend to reduce the glutathione level in the body if the interested person is aged enough or is not so well. Therefore, regular consumption of fruits and vegetables containing natural inflammatory compounds is sometimes questioned, and precautionary consumption measures are worth considering. It appears that natural inflammatory molecules contents in fruits and vegetables might act as safety parameters to decide whether a particular commodity is worth consuming or not in aged and ill conditions such as gout, obesity, hypertension, and diabetes.

2.3.1 Fibre and Fructose Contents of Popular Fruits

It is a common observation that fruits with high-fibre and/or low-fructose contents are safer than those with low-fibre and/or high-fructose contents [14]. Local and seasonal commodity with the preference of higher fibre and lower fructose contents is nowadays considered a suitable safety measure for selection of fruits for human consumption.

Fresh fruits containing more than 8% fructose might prove unsafe in accelerated oxidative stress conditions of the human body [14]. The safest fruits for regular consumption are perhaps those which contain fructose less than 5% and fibre greater than 15%. The safest fruits, suitable for regular consumption perhaps are as follows: kiwi fruit, peach, orange, strawberry, plum, lemon, pear, blueberry, raspberry, gooseberry, and cranberry.

It is said that people who have fructose intolerance—abdominal pain, diarrhoea, and gas—should limit high-fructose foods, such as juices, apples, grapes, watermelon, asparagus, pear, and *zucchini*; the lower fructose foods—such as bananas, blueberries, strawberries carrots, avocados, green bean, and lettuce—may be tolerated in limited quantities with meals [15]. Needless to say, such people should surely avoid fructose, high-fructose corn syrup, honey, agave syrup, invert sugar, maple-flavoured syrup, molasses, palm or coconut sugar, and sorghum [16]. Dried fruits such as dates, figs, and raisins obviously contain higher amounts of sugars, including fructose. It is also said that bananas and mangoes are equally high in fructose, but mangoes have less glucose. Consequently, they usually cause more problems; low-fructose and high-fibre foods are a good help for intestinal functions [17].

2.3.2 Low Purines Content of Popular Vegetables

The vegetarian diet is quite safer than the non-vegetarian diet with the viewpoint of the presence of purines, the gout-causing molecules, in the different foodstuffs.

In case, commodities containing purines content equal to or less than 50 mg per 100 g are arbitrarily considered for safety concerns, and three listed non-vegetarian foodstuffs, turkey escalope (raw), pork chop, and black pudding, stand as safe. However, all valuable and commonly used vegetarian foodstuffs prove to be quite safe. It is advised that daily food intake should create no more than 300 mg of uric acid [18]. It means the daily diet should not contain more than 131 mg of purines. This dietary guideline imposes a sort of restriction on the consumption of some widely used vegetables having moderate values of purines content (19 mg/100 g to 25 mg/100 g), namely mushrooms, Brussels sprouts, broccoli, spinach, and cauliflower, particularly in case of proneness to gout. Legumes are also restricted for the same reason.

Although alcoholic beverages contain little or no purine, they inhibit uric acid excretion and thus increase the uric acid level [18]. Milk and yoghurt (curd) possess negligible contents of purine (almost zero) and are worthy as purine-free foodstuffs.

2.3.3 Changes in Alkaloids Content of Vegetables During Processing

The inflammatory effect of glycoalkaloids is particularly worth mentioning in the case of nightshades such as potatoes, tomatoes, eggplants, and peppers. However, an increase in alkaloids content of vegetables, particularly leafy vegetables, during blanching with boiled water and juicing by means of juice extractor, is a matter of great caution. The juicing increases the alkaloids level dramatically in all the vegetables, perhaps due to increased synthesis of alkaloids occasioned by tempering heat of blanching and juicing. As a result, care must be exercised in consuming juice extract of vegetables [18].

2.4 Conclusions

It is concluded that

(a) The concept of high-fibre and low-fructose contents limits a few fruits for regular consumption; these fruits are kiwi fruit, peach, orange, strawberry, plum, lemon, pear, blueberry, raspberry, gooseberry, and cranberry.

(b) The requirement of a limiting value of purines, gout-causing compounds, less than 130 mg per day, imposes a restriction on several widely used vegetables: mushrooms, Brussels sprouts, broccoli, spinach, and cauliflower. However, these vegetables might be prepared using enough amounts of yoghurt (curd), a no purine foodstuff.

(c) The alkaloids content of vegetables, particularly leafy vegetables, is increased by the processes of blanching and juicing perhaps due to thermal interaction leading to increased synthesis of inflammatory alkaloid molecules.

(d) Finally, glycoalkaloids present as natural inflammatory molecules in night-shades such as potatoes, tomatoes, eggplants, and peppers are worth considering.

References

1. Boyce A, Jenking CM (1980) Autotrophic nutrition. In: Metabolism, movement and control. Foundations of Biology. Palgrave, London. https://doi.org/10.1007/978-1-349-04705-5_5
2. Anonymous (2021) Sources of sugar. The Canadian Sugar Institute, Toronto. https://sugar.ca/sugar-basics/sources-of-sugar. Accessed 25 June 2021
3. Stal LJ (2015) Nitrogen fixation in cyanobacteria. Encycl Life Sci (eLS) 2015:1–9. https://doi.org/10.1002/9780470015902a0021159.pub2
4. Hafez RM, Naguib RM (2017) Uric acid in plants and microorganisms: biological applications and genetics—a review. J Adv Res 8(5):475–486. https://doi.org/10.1016/j.jare.2017.05.003
5. Nakagawa T, Lanaspa M, Johnson R (2019) The effects of fruit consumption in patients with hyperuricemia or gout. Rheumatol 58(7):1133–1141. https://doi.org/10.1093/rheumatology/kez128
6. Cheong T (2021) Gout diet: top carbohydrates, dairy, fruits, and vegetables to eat. Health Xchange. https://www.healthxchange.sg/food-nutrition/food-tips/gout-diet-carbohydrate-dairy-fruit-vegetable. Accessed 25 June 2021
7. Chen C, Lin L (2020) Alkaloids in diet. In: Xiao J, Sarker SD, Asakawa Y (eds) Handbook of dietary phytochemicals. Springer Nature Singapore Pte Ltd, pp 1.35. https://doi.org/10.1007/978-981-13-1745-3_36-1
8. Skerrett PJ (2011) Is fructose bad for you? Harvard Health Publishing, Harvard. https://www.health.harvard.edu/blog/is-fructose-bad-for-you-200705012507. Accessed 25 June 2021
9. Kovacs Harbolic B (1996) What are foods that cause gout to flare up? MedicineNet. Inc. https://www.medicinenet.com/gout_foods_help_hurt_diet/article.htm. Accessed 25 June 2021
10. Nave CR (2017) Antenna complexes for photosynthesis. Photon absorption for photosynthesis. HyperPhysics, Department of Physics and Astronomy, Georgia State University. http://hyperphysics.phy-astr.gsu.edu/hbase/Biology/antpho.html. Accessed 25 June 2021
11. Finar IL (1959) Carbohydrates (Hexoses $C_6H_{12}O_6$). In: Organic chemistry, volume 1: The fundamental principles. English Language Book Society/Longman, pp 504–516
12. Frothingham S (2018) What is relationship between gout and sugar? Healthline. Updated on 29 October, 2018. www.healthline.com. https://www.healthline.com/health/gout-and-sugar. Accessed 25 June 2021
13. Stasolla C, Katahira R, Thorpe TA, Ashihara H (2003) Purine and pyrimidine nucleotide metabolism in higher plants. J Plant Physiol 160(11):1271–1295. https://doi.org/10.1078/0176-1617-01169
14. Anonymous (2021) No Fructose_Fruits. NoFructose.com. http://www.nofructose.com/food-ideas/fruit/. Accessed 25 June 2021
15. Zeratsky K (2019) Fructose intolerance: which foods to avoid? Mayo Clinic, Rochester. https://www.mayoclinic.org/fructose-intolerance/expert-answers/faq-20058097. Accessed 25 June 2021
16. Anonymous (2014) Low fructose diet. University of Virginia Health System, Charlottesville. https://med.virginia.edu/ginutrition/wp-content/uploads/sites/199/2014/04/Low-Fructose.pdf. Accessed 25 June 2021
17. Hendel B (2021) Purine content of foodstuffs. dr-barbara-hendel.com. Available https://dr-barbara-hendel.com/en/nutrition/tables/purine-content-table/. Accessed 25 June 2021
18. Odufuwa KT, Daramola GG, Adeniji PO, Salau BA (2013) Changes in alkaloids content of some selected Nigerian vegetables during processing. IOSR J Dent Med Sci 6(1):51–54

Chapter 3
The Role of Glycoalkaloids, Lipids, and Proteins in Tissue Inflammation

Abstract This chapter discusses the importance of food glycoalkaloids, the naturally produced nitrogen-containing compounds occurring as steroidal glycosides, as secondary metabolites produced in various parts of *Solanaceae* vegetables. Despite possessing appreciable amounts of vitamins and minerals, and being known for several health benefits, nightshade vegetables can create problems for people who are allergic to the alkaloids. The adverse health effects due to consumption of nightshade vegetables have been in limelight since 1979 when an epidemic of potato poisoning in London was reported. However, Ayurvedic physicians in India had been advising every patient to avoid nightshade vegetables consumption for centuries with an understanding that potatoes, tomatoes, eggplants, and peppers can generate excess heat and acidity in the body, aggravating both *vata* and *pitta* doshas leading to gout, nowadays called rheumatoid arthritis, high-level inflammation, and autoimmune diseases. In this chapter, an attempt is made to present a scenario of nutritional and detrimental effects of different nightshade vegetables, along with a mention of scientific opinion on the risk assessment of glycoalkaloids in feed and food. Finally, the Chapter concludes by highlighting the logic of dietary prohibitions suggested by Ayurvedic physicians to patients, and some introduction to the importance of proteins and lipids in tissue inflammation.

Keywords *Ayurveda* · Food safety · In vitro · In vivo · Oxidation · Phenolics · Secondary metabolite

Abbreviations

EC	European Commission
EFSA	European Food Safety Authority
GA	Glycoalkaloid
IBD	Inflammatory bowel disease
TPGA	Total potato glycoalkaloid

3.1 Introduction to Glycoalkaloids

Glycoalkaloids are a group of nitrogen-containing compounds, naturally produced in various cultivated and ornamental plant species of the *Solanaceae* family, including commonly consumed vegetables such as potatoes, tomatoes, eggplants, and peppers [1]. These compounds, occurring as steroidal glycosides, are secondary metabolites able to fight against pathogens, herbivores, and other adverse environmental stimuli in the leaves, flowers, roots, and edible parts including sprouts and skin of the plants of *Solanaceae* family. These species- or genera-specific molecular identities improve plant viability by increasing their overall ability to cope with local environmental changes, and are bitter and/or toxic to potential herbivores, often affecting the central and peripheral nervous system [2]. Although some nightshade vegetable species are toxic, most are standard agricultural crops, possessing appreciable amounts of vitamins and minerals necessary for a healthy diet [3].

For example, white potatoes contain more potassium, needed to recover from workouts and keep muscles properly functioning, than bananas; tomatoes are rich in immune system boosting vitamin C, and contain serum lipid oxidation preventing molecule lycopene [3].

Despite being known for several health benefits, nightshade vegetables are advised to be avoided for people who are allergic to alkaloids and may experience one or more of the following symptoms after eating a vegetable from the nightshade family: hives and skin rashes, itchiness, nausea, vomiting, excessive mucus production, achy muscles and joints, and inflammation [4]. The adverse health effects due to consumption of nightshade vegetables have been in limelight since 1979 when McMillan and Thomson reported an epidemic of potato poisoning among 78 schoolboys in London after having eaten lunch with prepared potatoes containing 250–300 mg/kg of glycoalkaloids [5].

In addition, it would be worth mentioning that Ayurvedic physicians, called *vaidyas*, had been prohibiting nightshades—tomatoes, eggplants, potatoes, and chili—and any sour-tasting food substance, jaggary, and vegetable oil (of course allowing milk and milk fat called ghrut or ghee) in India to every patient, almost irrespective of disease whether diagnosed by them as *vata* or *pitta* or *kapha*, since centuries. From an Ayurvedic perspective, nightshades—white potatoes, tomatoes, peppers, eggplants, bell peppers, cayenne pepper, paprika, goji berries, jalapenos, and pimentos—possess inflammatory properties which can generate excess heat and acidity in the body aggravating both *vata* and *pitta* doshas (ailments) which nowadays are diagnosed as rheumatoid arthritis, high levels of inflammation, and autoimmune diseases [6].

In this chapter, an attempt is made to present a scenario of both the nutritional and the detrimental effects of nightshade vegetables on human health. In this ambit, the chemical configuration of glycoalkaloid molecules, occurring in different night-shades, and their concentration in different plant parts are mentioned to underline the presumed reason why nightshades may not agree with the human body system. The chapter also mentions a scientific opinion on the risk assessment of glycoalkaloids in feed and food, in particular, in potato and potato-derived products [7]. The Chapter finally concludes that dietary prohibitions suggested by Ayurvedic physicians to patients appear to be logical.

3.2 Nutritional Effects of Nightshade Vegetables to Human Health

Despite some of the nightshades possessing detrimental effects on human health, the *Solanaceae* family of plants covering almost two thousand species—characterised by distinctive features like the shape of the flower and the arrangement of seeds within the fruit—is considerably known for nutritional values.

For example, potatoes or *Solanum tuberosum* (barring sweet potatoes, i.e *Ipomoea batatas*) contain considerable amount of potassium, vitamin B_6, and manganese; tomatoes or *S. lycopersicum* are full of vitamin A and C with anticarcinogenic and antioxidant lycopene; eggplants or *S. melongena* are potent weight-loss agents because they contain 2.5 g of fibre per cup; nell peppers (sweet peppers) or *Capsicum annuum*, in the bell shape-fruit group, are a storehouse of vitamin C and possess an excellent capability of iron absorption. Moreover, chili peppers like *bhut Jolokia* (hybrid of *C. frutescens* and *C. chinense*), Indian pepper or Chile tepon, Turkey or bird's eye (*C. annuum* var. *glabriusculum*), habaneros (*C. chinense*), *peri peri* (*C. frutescens*), cayenne, and jalapenos (both *C. annuum*) contain a heat level-boosting component called capsaicin, helpful in reducing calorie intake [8].

3.3 Detrimental Effects of Nightshade Vegetables on Human Health

Some of the *Solanaceae* nightshade foods—fruits and vegetables—are well known for Ayurvedic medicinal applications like *Ashwagandha* or *Withania somnifera* for rejuvenation, stress elimination, and neuroregeneration [9], goji berries or *Lycium barbarum* for improving kidney and liver function and providing anti-ageing properties that benefit skin [10], paprika and cayenne for enhancing circulation and pacifying *kapha* and *vata* [11]. Still, there is controversy regarding the health aspects of vastly consumed nightshade vegetables, with particular concern to potatoes, tomatoes, eggplants, and peppers, whether those are good or bad. These high nutritional-profile nightshades are usually prohibited by Ayurvedic physicians to the patients.

Of course, food prohibition for patients is normally not practiced by allopathic physicians. On the other side, a common perception nowadays exists: the regular consumption of potatoes, tomatoes, eggplants, and peppers might lead to inflammation in the body, and suffering from autoimmune diseases is large because of nightshades [8]. The set of diseases presumed to be caused by nightshade vegetables primarily includes inflammatory bowel disease (IBD), Crohn's disease, and gastrointestinal problems such as diarrhoea [8]. It is also now believed that increased intestinal permeability or leaky gut can give rise to autoimmune conditions like celiac disease, multiple sclerosis, and rheumatoid arthritis [8]. Some of the people are now said to be particularly sensitive to nightshades and they suffer from conditions like allergy system (hives, skin rashes, swelling, itching in throat), who feel relief after elimination of nightshades from their diet [8]. These nightshade vegetables contain glycoalkaloid molecules, obviously forming the basis of presumed reasons for why nightshades may not agree with the human body system, and sometimes yield detrimental effects.

3.4 Chemical Configurations of Much Discussed Nightshade Glycoalkaloid Molecules

Despite some medicinal applications of glycoalkaloid molecules present in *Solanaceae* plants,—potatoes, tomatoes, eggplants, and peppers—these molecules are nowadays considered responsible for inflammation and allergic conditions in the human body on regular consumption. The red and green bell pepper fruits are exceptional in this context because they contain very small amount of glycoalkaloids—less than 10 mg per kg—to cause inflammation or allergic symptoms [12]. Should the pepper sensitivity be demonstrated, the person would be responding to other compounds containing in peppers, such as the hot and spicy capsaicinoids [12]. Therefore, much discussed glycoalkaloids—responsible for inflammation and allergic symptoms—are found in potatoes (α-solanine and α-chaconine), aubergine, i.e. eggplants (α-solasonine and α-solamargine), and tomatoes (α-tomatine and α-dehydrotomatine) [7].

Glycoalkaloids are composed of a steroidal aglycone part and an oligosaccharide side chain attached to the 3-hydroxyl group of the aglycone; most glycoalkaloids (GA) contain either a trisaccharide (chacotriose or solatriose) or a tetrasaccharide (lycotetraose) [7]. In commercial potato cultivars (*S. tuberosum*), mostly the glycoalkaloids α-solanine and α-chaconine are composed of a solanidine aglycone with solatriose and chacotriose, respectively.

The aubergine fruit (derived from *S. melogena*) contains primarily glycoalkaloids α-solasonine and α-solamargine which are composed of solasodine aglycone with solatriose and chacotriose, respectively. In tomato fruit (derived from *S. lycopersicum* varieties), α-tomatine and α-dehydrotomatine are the major glycoalkaloid molecules, composed of the aglycones tomatidine and tomatidinol, respectively, coupled to lycotetraose [7].

3.5 Glycoalkaloid Concentrations in Different Plant Parts

As far as the GA content of edible (for humans) plant parts, that means potato tubers and tomato and eggplant fruits are concerned, these differ as per taste, maturity, and development factors. The potato tubers with skin normally possess a total GA concentration of 75 mg per kg of fresh weight if the taste is quite normal; however, it is found in the range 250–800 mg/kg fresh weight for bitter tubers with skin, while peel (skin) and tuber without peel contain 150–600 mg/kg and 12–50 mg/kg of GA, respectively [13].

GA concentration for tomato fruit depends on its maturity and effectively decreases with an increase in maturity or ripening. The small immature green tomato fruit contains approximately 500 mg/kg of GA, while large immature green fruit possesses around 150 mg/kg of GA; on the other side, tomatine and dehydrotomatine are largely degraded when the fruit ripens, to a level of only 5 mg/kg of fresh fruit weight in red tomatoes [13].

On the contrary, GA content in eggplant fruits generally increases with development and ripening; a calorimetric study of 21 different varieties of *S. melongena* as carried out by Bajaj and coworkers showed GA content ranged from 62.5 to 205 mg/kg of fresh weight (mean value: 113 mg/kg) [13].

GA concentration for potato sprouts, flowers and leaves are very high ranging 2,000–4,000, 3,000–5,000, and 400–1,500 mg/kg, respectively. However, this amount is just around 30 mg/kg for stem [13]. On the other side, the GA content of potato leaves is around 1,000 mg/kg, and this value for senescent leaves is greater than 5,000 mg/kg. The average GA concentration for tomato roots, calyxes, stems, and flowers is around 150, 850, 800, and 130 mg/kg, respectively [13].

3.6 Presumed Reason for Adverse Human Body Reactions to Nightshades Consumption

The nightshade glycoalkaloids are known for both medicinal and detrimental values. Despite exhibiting a number of interesting and potentially useful interactions with cell membranes in their reported roles such as anticancer agents, antifungal agents, and in vaccine as adjuvants [14], the glycoalkaloids, particularly contained in potatoes, may induce gastrointestinal and systemic effects by cell membrane disruption and acetylcholinesterase inhibition [15]. The twenty-first century researches particularly conducted by D. S. McGehee and coworkers supported the hypothesis that inhibition of endogenous enzyme systems by dietary factors can influence anaesthetic drug metabolism and duration of action (hospital anaesthesia warning) because cholinesterase inhibition by potato glycoalkaloids slows mivacurium metabolism [16]. Acetylcholinesterase inhibition induced by glycoalkaloids may be regarded as a presumed reason why nightshades may not agree with the human body system.

Acetylcholinesterase is involved in the termination of impulse transmission by rapid hydrolysis of the neurotransmitter acetylcholine in numerous cholinergic pathways in the central and peripheral nervous system. The enzymatic inactivation, induced by various inhibitors (say nightshade glycoalkaloids), leads to acetylcholine accumulation, hyperstimulation of nicotinic and muscarinic receptors, and disrupted neurotransmission [17], obviously leading to oxidative stress and inflammation conditions.

3.7 A Scientific Opinion on the Risk Assessment of Glycoalkaloids in Food

The European Food Safety Authority (EFSA) has recently responded to the European Commission (EC) with a scientific opinion on the risk assessment of glycoalkaloids in feed and food, particularly in potatoes and potato-derived products. On the basis of available data, it has been clearly opined that in humans potato GA are systemically absorbed following ingestion; acute toxic effects following ingestion of total potato glycoalkaloids (TPGA) include gastrointestinal symptoms of varying severity such as vomiting, diarrhoea, and abdominal pain, which may occur from a TPGA intake of 1 mg/kg body weight or more [7]. Further, (after slightly more intake) symptoms including drowsiness, apathy, confusion, weakness, vision disturbances, rapid and weak pulse, and low blood pressure may be the consequence of dehydration following vomiting and diarrhoea [7].

TPGA 3–6 mg/kg body weight is considered to be potentially lethal for humans when speaking of TPGA intake. The Scientific Opinion report states that there are no toxicokinetic data on tomato and aubergine GA (and their aglycones) in experimental animals and humans, and identifies needs of occurrence date [7].

3.8 Dietary Prohibition of Ayurvedic Physicians to Patients Appears Logical

The GA concentration in potato tuber of normal taste (not bitter) without skin is in the range 12–50 mg/kg [13]. It appears to be quite safe for human consumption, keeping in mind the safety concerns demanding less than 1mg/kg body weight [7]. However, dried, baked, or fried potatoes (potato chips) might be very harmful, if regularly consumed in considerably high amounts. The other parts of the potato plant such as sprouts, flowers, and leaves should sure be avoided. Considering GA concentration in green and ripe red tomatoes, the selection of the latter appears to be logical; the same logic proves the immature small-sized eggplants safer than the developed big fruits [13]. However, the EFSA's Scientific Opinion on risk assessment of glycoalkaloids in feed and food [7] persuades people to accept the fact that dietary prohibitions regarding nightshades of Ayurvedic physicians to patients is perhaps logical.

3.9 The Problem of Lipids and Proteins in Tissue Inflammation. An Introduction

The discussion concerning the risk assessment of GA in feed and food products can be useful when speaking of general target molecules for oxidative attacks. As above-mentioned, GA are a group of nitrogen-containing compounds, naturally found in the *Solanaceae* family (potatoes, tomatoes, etc.). The existence of other nitrogen-containing molecules—general proteins which might be considered as soft targets for oxidative attack—has to be mentioned. In addition, the role of *Solanaceae*-derived ingredients and additives for the production of high-energy foods (Chap. 5) and other products containing remarkable amounts of oxidation-sensible lipids should be remembered.

Substantially, the presence of food contaminants—chlorine dioxide in flour; benzoyl peroxide in bread; etc.—able to oxidise lipids (various fats and oils) and long-chain nitrogen-based molecules (proteins) should be taken into account when speaking of public health perspectives.

First of all, it should always be remembered that food success or palatability—in sensorial terms—is directly proportional to the quantity of sweet carbohydrates (sugars) and lipids. The synergetic effect of these fats/oils and added sugars is also well known and studied. On the other side, the oxidation of added fats and oils may be partially avoided if some antioxidant agents of natural origin are added to the intermediate food. In this way, the oxidative attack could be partially limited, depending on the 'right choice' of natural additives and related quantities. More research is needed in this ambit, also with respect to the effect of fatty acids on lipid peroxidation and the production of antioxidant enzymatic agents [18–34].

On the other side, vegetable fats/oils and vegetable proteins, in form of processed ingredients, can be used in the production of ultraprocessed foods. These products should be considered as a concentrated mass of high-energy chemical compounds (lipids, protein, and sugars). Actually, the market can also show 'light' ultraprocessed foods with reduced trans fat contents, low sugars, and so on. However, these products are compared with the original version of ultraprocessed foods. Consequently, the comparison may be questionable by the health viewpoint at least (Chap. 5).

In the ambit of products containing high amounts of lipids and protein, it should be always considered that these nutrients are sensible to oxidation, and this phenomenon can be remarkably accelerated by means of processing and storage operations in the food industry. This problem can be a real concern when speaking of vegan foods because they can contain remarkable amounts of vegetable (sunflower, palm, and coconut) oils, and/or protein. The example of cheese-like vegan products is interesting in this ambit because their composition relies on starch and vegetable (sunflower, palm, and coconut) oils, while proteins are not generally considered [35]. As a result, the oxidative resistance of these molecules could be a concern in the ambit of public safety. In fact, their demolition is often correlated with one or more of the following diseases (Chap. 1):

(a) Atherosclerosis
(b) Cardiovascular diseases, including hypertension
(c) Diabetes
(d) Dysfunction of the respiratory system, including asthma
(e) Neurological disorders
(f) Tumours.

For these and other reasons, more research is needed—in spite of the remarkable amount of research papers—concerning the protection of lipids in the ambit of foods and beverages [36–43]. The same thing can be affirmed when speaking of sensible protein molecules [44–51]. Chapter 5 gives a discussion concerning the importance of these oxidation targets, with reference to:

(1) Polyunsaturated fatty acids when speaking of lipids (and related reactions: lipid peroxidation), and
(2) Proteins that could be able to react with toxic aldehydes (obtained by lipid peroxidation and by means of other reaction pathways), with resulting covalent bonds between two amino groups of two different chains, and the consequent production of Schiff bases and possible heterocyclic compounds.

In relation to protein molecules, these chains can undergo three different oxidation attacks: breaking of peptide bonds, oxidation of amino acid residues, and covalent links between different proteins. These risks have to be taken into account with other visible (and worrying) phenomena: Maillard reaction products and caramelisation above all.

References

1. Chen C, Lin L (2020) Alkaloids in diet. In: Xiao J, Sarker S, Asakawa Y (eds) Handbook of dietary phytochemicals. Springer, Singapore. https://doi.org/10.1007/978-981-13-1745-3_36-1
2. Kennedy DO, Wightman EL (2011) Herbal extracts and phytochemicals: plant secondary metabolites and the enhancement of human brain functions. Adv Nutr 2(1):32–50. https://doi.org/10.3945/an.110.000117
3. Purdie J (2020) Nightshade vegetables list. Types of nightshade vegetable. Very Well Fit, Dotdash, Inc., New York. https://www.verywellfit.com/the-health-benefits-of-nightshade-vegetables-4687184. Accessed 02 Aug 2021
4. Lillis C (2018). What to know about nightshade allergies. Medical News Today, Healthline Media UK Ltd., Cheltenham. https://www.medicalnewstoday.com/articles/321883. Accessed 02 Aug 2021
5. McMillan M, Thomson JC (1979) An outbreak of suspected solanine poisoning in schoolboys: examinations of criteria of solanine poisoning. QJM Int J Med 48(2):227–243. https://doi.org/10.1093/oxfordjournals.qjmed.a067573

6. Gunthor KL (2021) Nightshades and Ayurveda. Why are nightshades avoided in Ayurveda? Lakshmi Ayurveda. www.lakhmiayurveda.com.qu. https://www.lakshmiayurveda.com.au/2021/03/nightshades-and-ayurveda/. Accessed 02 Aug 2021
7. EFSA Panel on Contaminants in the Food Chain (CONTAM), Schrenk D, Bignami M, Bodin L, Chipman JK, del Mazo J, Hogstrand C, Hoogenboom LR, Leblanc J-C, Nebbia CS, Nielsen E, Ntzani E, Petersen A, Sand S, Schwerdtle T, Vleminckx C, Wallace H, Brimer L, Cottrill B, Dusemund B, Mulder P, Vollmer G, Binaglia M, Ramos Bordajandi L, Riolo F, Roldan-Torres R, Grasl-Kraupp B (2020) Risk assessment of glycoalkaloids in feed and food, in particular in potatoes and potato-derived products. EFSA J 18(8):e06222. https://doi.org/10.2903/j.efs a2020.6222
8. Devi G (2020) Guide on nightshade vegetables and fruits in your diet—best replacement foods. The Fit Indian. www.thefitindian.com. https://www.thefitindian.com/blog/nightshade-food-risks/. Accessed 02 Aug 2021
9. Singh N, Bhalla M, de Jager P, Gilea M (2011) An overview on Ashwagandha: a rasayan (rejuvenator) of Ayurveda. Afr J Trad Compl Altern Med 8(5):S208–S213. https://doi.org/10.4314/ajtcam.v8i5S.9
10. Anonymous (2021) Goji Berries: the Ayurvedic ingredient that can boost your skincare routine. AVYA Advanced Ayurvedic Skincare (AVYA). www.avyaskincare.com. https://www.avyaskincare.com/blogs/blog/goji-berries-the-ayurvedic-ingredient-that-can-boost-your-ski ncare-routine. Accessed 02 Aug 2021
11. Maharishi Ayurveda Staff (2021) Sweet Paprika. Maharishi AyurVeda Products International, Inc., Fairfield. www.mapi.com. https://mapi.com/blogs/articles/sweet-paprika. Accessed 02 Aug 2021
12. Ede G (2021) How deadly are nightshades? Diagnosis: DIET. www.diagnosisdiet.com. https://www.diagnosisdiet.com/full-article/nightshades. Accessed 02 Aug 2021
13. Siddique MAB, Brunton N (2019) Food Glycoalkaloids: distribution, structure, cytotoxicity, extraction, and biological activity. In: Kurek J (ed) Alkaloids—their importance in nature and human life. IntechOpen, London. https://doi.org/10.5772/intechopen.82780
14. Nepal B, Stine KJ (2019) Glycoalkaloids: structure, properties, and interaction with model membrane system. Processes 7(8):513. https://doi.org/10.3390/pr7080513
15. Mensinga TT, Sips AJAM, Rompelberg CJM, van Twillert K, Meulenbelt J, van den Top HJ, van Egmond HP (2005) Potato glycoalkaloids and adverse effects in humans: an ascending dose study (clinical trial). Reg Toxicol Pharmacol 41(1):66–72. https://doi.org/10.1016/j.yrtph.2004.09.004
16. McGehee DS, Krasowski MD, Fung DL, Wilson B, Gronert GA, Moss J (2000) Cholinesterase inhibition by potato glycoalkaloids slows mivacurium metabolism. Anesthesiology 93:510–519. https://doi.org/10.1097/00000542-200008000-00031
17. Colovic MB, Krstic DZ, Lazarevic-Pasti TD, Bondzic AM, Vasic VM (2013) Acetyl-cholinesterase inhibitors: pharmacology and toxicology. Curr Neuropharmacol 11(3):315–335. https://doi.org/10.2174/157159X1311030006
18. Ruberto G, Baratta MT (2000) Antioxidant activity of selected essential oil components in two lipid model systems. Food Chem 69:167–174. https://doi.org/10.1016/S0308-8146(99)002 47-2
19. Yu JQ, Lei JC, Zhang XQ, Yu HD, Tian DZ, Liao ZX, Zou GL (2011) Anticancer, antioxidant and antimicrobial activities of the essential oil of Lycopus lucidus Turcz. var. hirtus Regel. Food Chem 126(4):1593–1598. https://doi.org/10.1016/j.foodchem.2010.12.027
20. Lee YJ, Kang DG, Kim JS, Lee HS (2008) Lycopus lucidus inhibits high glucose-induced vascular inflammation in human umbilical vein endothelial cells. Vasc Pharmacol 48(1):38–46. https://doi.org/10.1016/j.vph.2007.11.004
21. Lu YH, Huang JH, Li YC, Ma TT, Sang P, Wang WJ, Gao CY (2015) Variation in nutritional compositions, antioxidant activity and microstructure of Lycopus lucidus Turcz. root at different harvest times. Food Chem 183:91–100. https://doi.org/10.1016/j.foodchem.2015.03.033
22. Yang X, Lv Y, Tian L, Zhao Y (2010) Composition and systemic immune activity of the polysaccharides from an herbal tea (Lycopus lucidus Turcz). J Agric Food Chem 58(10):6075–6080. https://doi.org/10.1021/jf101061y

23. Freitas RM (2009) The evaluation of effects of lipoic acid on the lipid peroxidation, nitrite formation and antioxidant enzymes in the hippocampus of rats after pilocarpine-induced seizures. Neurosci Lett 455(2):140–144. https://doi.org/10.1016/j.neulet.2009.03.065

24. Militão GCG, Ferreira PMP, de Freitas RM (2010) Effects of lipoic acid on oxidative stress in rat striatum after pilocarpine-induced seizures. Neurochem Int 56(1):16–20. https://doi.org/10.1016/j.neuint.2009.08.009

25. Xavier SM, Barbosa CO, Barros DO, Silva RF, Oliveira AA, Freitas RM (2007) Vitamin C antioxidant effects in hippocampus of adult Wistar rats after seizures and status epilepticus induced by pilocarpine. Neurosci Lett 420(1):76–79. https://doi.org/10.1016/j.neulet.2007.04.056

26. Ferreira PMP, Militão GCG, Freitas RM (2009) Lipoic acid effects on lipid peroxidation level, superoxide dismutase activity and monoamines concentration in rat hippocampus. Neurosci Lett 464(2):131–134. https://doi.org/10.1016/j.neulet.2009.08.051

27. dos Santos Sales ÍM, Do Nascimento KG, Feitosa CM, Saldanha GB, Feng D, de Freitas RM (2011) Caffeic acid effects on oxidative stress in rat hippocampus after pilocarpine-induced seizures. Neurol Sci 32(3):375–380. https://doi.org/10.1007/s10072-010-0420-4

28. Tome AR, Feng D, Freitas RM (2010) The effects of alpha-tocopherol on hippocampal oxidative stress prior to in pilocarpine-induced seizures. Neurochem Res 35(4):580–587. https://doi.org/10.1007/s11064-009-0102-x

29. Schwarz K, Bertelsen G, Nissen LR, Gardner PT, Heinonen MI, Hopia A, Huynh-Ba T, Lambelt P, McPhail D, Skibsted LH, Tijburg L (2001) Investigation of plant extracts for the protection of processed food against lipid oxidation. Comparison of antioxidant assays based on radical scavenging, lipid oxidation and analysis of the principal antioxidant compounds. Eur Food Res Technol 212:319–328. https://doi.org/10.1007/s002170000256

30. Lebeau J, Furman C, Bernier JL, Duriez P, Teissier E, Cotelle N (2000) Antioxidant properties of di-tert-butylhydroxylated flavonoids. Free Rad Biol Med 29(9):900–912. https://doi.org/10.1016/S0891-5849(00)00390-7

31. Gardner PT, McPhail DB, Duthie GG (1998) Electron spin resonance spectroscopic assessment of the antioxidant potential of teas in aqueous and organic media. J Sci Food Agric 76(2):257–262. https://doi.org/10.1002/(SICI)1097-0010(199802)76:2%3C257::AID-JSFA944%3E3.0.CO;2-B

32. Møller JK, Madsen HL, Aaltonen T, Skibsted LH (1999) Dittany (Origanum dictamnus) as a source of water-extractable antioxidants. Food Chem 64(2):215–219. https://doi.org/10.1016/S0308-8146(98)00143-5

33. Ramadan MF, Kroh LW, Mörsel JT (2003) Radical scavenging activity of black cumin (Nigella sativa L.), coriander (Coriandrum sativum L.), and niger (Guizotia abyssinica Cass.) crude seed oils and oil fractions. J Agric Food Chem 51(24):6961–6969. https://doi.org/10.1021/jf0346713

34. Agwaramgbo L, Okegbe T, Wright T, Igwe S, Ogburie V (2013) Inhibition of the oxidation of acetophenone by aqueous extracts of edible fruits and vegetables. Mod Chem Appl 1(3):1000105. https://doi.org/10.4172/2329-6798.1000105

35. Haddad MA, Omar SS, Parisi S (2021) Vegan cheeses vs processed cheeses—traceability issues and monitoring countermeasures. Br Food J 123(6):2003–2015. https://doi.org/10.1108/BFJ-10-2020-0934

36. Bou R, Navas JA, Tres A, Codony R, Guardiola F (2012) Quality assessment of frying fats and fried snacks during continuous deep-fat frying at different large-scale producers. Food Control 27(1):254–267. https://doi.org/10.1016/j.foodcont.2012.03.026

37. Rutkowska J, Antoniewska A, Martinez-Pineda M, Nawirska-Olszańska A, Zbikowska A, Baranowski D (2020) Black chokeberry fruit polyphenols: a valuable addition to reduce lipid oxidation of muffins containing xylitol. Antioxidants 9(5):394. https://doi.org/10.3390/antiox9050394

38. Kong J, Perkins LB, Dougherty MP, Camire ME (2011) Control of lipid oxidation in extruded salmon jerky snacks. J Food Sci 76(1):C8–C13. https://doi.org/10.1111/j.1750-3841.2010.01896.x

39. Bekele EK, Nosworthy MG, Henry CJ, Shand PJ, Tyler RT (2020) Oxidative stability of direct-expanded chickpea–sorghum snacks. Food Sci Nutr 8(8):4340–4351. https://doi.org/10.1002/fsn3.1731

40. Albertos I, Martin-Diana AB, Jaime I, Diez AM, Rico D (2016) Protective role of vacuum vs. atmospheric frying on PUFA balance and lipid oxidation. Innov Food Sci Emerg Technol 36:336–342. https://doi.org/10.1016/j.ifset.2016.07.006

41. EFSA Panel on Dietetic Products Nutrition and Allergies (2011) Scientific opinion on the substantiation of health claims related to soy isoflavones and protection of DNA, proteins and lipids from oxidative damage (ID 1286, 4245), maintenance of normal blood LDL cholesterol concentrations (ID 1135, 1704a, 3093a), reduction of vasomotor symptoms associated with menopause (ID 1654, 1704b, 2140, 3093b, 3154, 3590), maintenance of normal skin tonicity (ID 1704a), contribution to normal hair growth (ID 1704a, 4254), "cardiovascular health" (ID 3587), treatment of prostate cancer (ID 3588) and "upper respiratory tract" (ID 3589) pursuant to Article 13(1) of Regulation (EC) No 1924/2006. EFSA J 9(7):2264–2308. https://doi.org/10.2903/j.efsa.2011.2264

42. Couet C, Delarue J, Ritz P, Antoine JM, Lamisse F (1997) Effect of dietary fish oil on body fat mass and basal fat oxidation in healthy adults. Int J Obes 21(8):637–643. https://doi.org/10.1038/sj.ijo.0800451

43. Difonzo G, Pasqualone A, Silletti R, Cosmai L, Summo C, Paradiso VM, Caponio F (2018) Use of olive leaf extract to reduce lipid oxidation of baked snacks. Food Res Int 108:48–56. https://doi.org/10.1016/j.foodres.2018.03.034

44. Hellwig M (2020) Analysis of protein oxidation in food and feed products. J Agric Food Chem 68(46):12870–12885. https://doi.org/10.1021/acs.jafc.0c00711

45. Baskol M, Baskol G, Koçer D, Ozbakir O, Yucesoy M (2008) Advanced oxidation protein products: a novel marker of oxidative stress in ulcerative colitis. J Clin Gastroenterol 42(6):687–691. https://doi.org/10.1097/MCG.0b013e318074f91f

46. Zhong ZM, Bai L, Chen JT (2009) Advanced oxidation protein products inhibit proliferation and differentiation of rat osteoblast-like cells via NF-κB pathway. Cell Physiol Biochem 24(1–2):105–114. https://doi.org/10.1159/000227818

47. Apak R, Ozyurek M, Guclu K, Capanoglu E (2016) Antioxidant activity/capacity measurement. 3. Reactive oxygen and nitrogen species (ROS/RNS) scavenging assays, oxidative stress biomarkers, and chromatographic/chemometric assays. J Agric Food Chem 64(5):1046–1070. https://doi.org/10.1021/acs.jafc.5b04744

48. Estévez M, Li Z, Soladoye OP, Van-Hecke T (2017) Health risks of food oxidation. Adv Food Nutr Res 82:45–81. https://doi.org/10.1016/bs.afnr.2016.12.005

49. Zheng P, Bai X, Long J, Li K, Xu H (2016) Nitric oxide enhances the nitrate stress tolerance of spinach by scavenging ROS and RNS. Sci Hortic 213:24–33. https://doi.org/10.1016/j.scienta.2016.10.008

50. Falowo AB, Fayemi PO, Muchenje V (2014) Natural antioxidants against lipid–protein oxidative deterioration in meat and meat products: a review. Food Res Int 64:171–181. https://doi.org/10.1016/j.foodres.2014.06.022

51. Estévez M, Luna C (2017) Dietary protein oxidation: a silent threat to human health? Crit Rev Food Sci Nutr 57(17):3781–3793. https://doi.org/10.1080/10408398.2016.1165182

Chapter 4
Determination of Inflammatory Molecules in Fruits and Vegetables

Abstract Different vegetable foods could unpredictably cause similar inflammatory reactions when speaking of human diets. Several unrefined carbohydrates and fresh oils, and also several pulses can be the cause of mild surplus oxidation. These situations appear in contrast with common knowledge speaking of foods and beverages with antioxidant or anti-inflammatory properties. For these and other reasons, vegan/vegetarian dietary lifestyles appear to be successful at present. However, intensive farming practices concerning fruits and vegetables worldwide have progressively caused a decline in the amount of protein, phosphorus, iron, riboflavin, and vitamin C in modern food products belonging to this ambit. As a result, organophosphorus pesticides are now detectable in fruits, vegetables, and wheat. Because of their inflammatory effects on the human body, and also considering their persistency in the environment, these molecules are of extreme interest. This worry is also present when speaking of chemical agents used for artificial ripening, with concern to fruits. The category of powdered juices can also contain different food additives with possible inflammatory, cytotoxic, and /or genotoxic effects. Several inflammatory molecules of natural origin exist: fructose, purines, and alkaloids. This chapter concerns the analytical detection of these chemical classes.

Keywords Alkaloid · Calcium carbide · Fructose · Inflammatory effect · OP · Pesticide · Purine

Abbreviations

C_aC_2	Calcium carbide
CAD	Charged aerosol detection
EDX	Elemental composition analysis
ELSD	Evaporative light scattering detection
FT	Fourier Transform
FT-IR	Fourier Transform infrared spectroscopy
GC–MS/MS	Gas chromatography/tandem mass spectrometry
GC	Gas chromatography

© The Author(s), under exclusive license to Springer Nature Switzerland AG 2022
R. K. Sharma et al., *Natural Inflammatory Molecules in Fruits and Vegetables*,
Chemistry of Foods, https://doi.org/10.1007/978-3-030-88473-4_4

HS-SPME	Headspace solid-phase microextraction
HPLC	High-performance liquid chromatography
LC-MS	Liquid chromatography
MS	Mass spectrometry
NIR	Near-Infrared
OP	Organophosphorus
PLS	Partial least-squares
PDA	Photodiode array
RID	Refractive index detection
SEM	Scanning electron microscope
MS/MS	Tandem mass spectrometry

4.1 An Introduction to the Analytical Detection of Inflammatory Molecules in Fruits and Vegetables

The consumption of certain food products is reported (Chap. 11) to be one of the reasons for chronic inflammation in the human body. In general, this situation is reported when speaking of fast-food products and food articles largely consumed in the Western world, as displayed in Fig. 4.1 [1]:

(1)   Bakery products (because of their amount of refined carbohydrates)
(2)   Fried foods (because of problems caused by rancid oils used for frying, and also because of frying techniques and incorrect frying procedures)
(3)   Sugar-sweetened beverages and soda (because of the well-known excess of sugars if used/consumed in high-dietary lifestyles, with important safety and public health concerns)
(4)   Red meat products (examples: steaks, burgers)
(5)   Processed meat (examples: sausages, hot dogs)
(6)   Fat emulsions and/or solids such as margarine, shortening, and lard (used as ingredients in the food industry).

On the other side, different vegetable foods could unpredictably (for normal consumers which are normally unaware of food-related illnesses from a health/safety and professional viewpoint) cause similar inflammatory reactions when speaking of human diets. In detail, certain (unrefined) carbohydrates and fresh oils, and also several pulses can be the cause of mild surplus oxidation. The above-mentioned situations appear in contrast with common knowledge speaking of foods and beverages with antioxidant or anti-inflammatory properties, as displayed in Fig. 4.2 [2–5]:

(a)   Cereals as whole grains and derived products (also bread). Examples: barley, brown rice, oatmeal, etc.
(b)   Coffee, green tea
(c)   Dark chocolate (in dark and red versions).

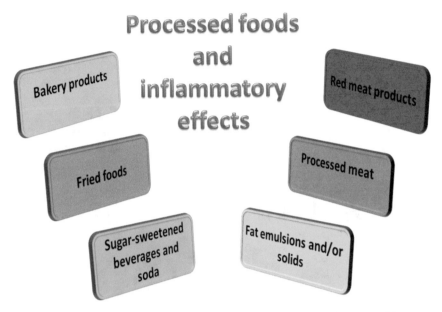

Fig. 4.1 The consumption of certain food products is reported (Chap. 1) to be as one of the reasons for chronic inflammation in the human body. This situation is reported when speaking of fast-food products and food articles largely consumed in the Western world [1]

(d)   Dried fruits. Example: prunes
(e)   Fatty fish. Examples: albacore, herring, lake trout, mackerel, salmon, etc.
(f)   Fresh fruits. Examples: apples, blueberries, grapefruit, mangoes, peaches, pomegranates, etc.
(g)   High-protein vegetable products (pulses). Examples: chickpeas, lentils, etc.
(h)   Leafy greens. Examples: romaine lettuce, spinach, etc.
(i)   Non-high protein vegetables. Examples: broccoli; Brussels sprouts, cauliflower, etc.
(j)   Nuts. Examples: almonds, walnuts.

For these and other reasons, vegan/vegetarian dietary lifestyles appear to be successful at present, because many of these anti-inflammatory products are substantially of vegetable origin [6].

On the other hand, the intensive farming practices concerning fruits and vegetables worldwide have progressively caused a decline in the amount of protein, phosphorus, iron, riboflavin (vitamin B_2), and vitamin C in modern food products belonging to this ambit [7]. As a result, organophosphorus (OP) pesticides such as chlorpyrifos, malathion, methamidophos, and malaoxon, are now detectable in fruits, vegetables, and wheat. Because of their toxic/inflammatory effects on human health, and also considering their persistency in the environment, these molecules are of extreme interest when speaking of artificial and environmental contamination, on the one side, and of public health on the other side. This worry is also present when speaking

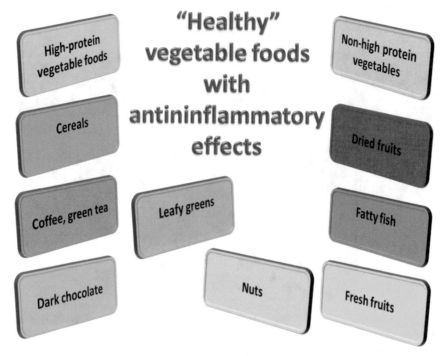

Fig. 4.2 Different vegetable foods could unpredictably cause similar inflammatory reactions when speaking of human diets. In detail, certain (unrefined) carbohydrates and fresh oils, and also several pulses can be cause of mild surplus oxidation. The above-shown situations appear in contrast with common knowledge speaking of foods and beverages with antioxidant or anti-inflammatory properties [2–5]

of chemical agents used for artificial ripening (with concern to fruits), such as calcium carbide (C_aC_2), in spite of reported inflammatory and toxic effects [8]. In addition, the category of powdered juices can contain different food additives with possible inflammatory/ cytotoxic/ genotoxic effects [9].

Moreover, the category of pesticides is believed to interfere with oxidation/reduction equilibrium in humans as a consequence of the modification of glutathione levels (glutathione is a natural antioxidant or free radical scavenger, with detoxifying properties, in vegetables). As a result, the occurrence of neurodegenerative diseases such as Alzheimer's and Parkinson's disease have been reported to be correlated with glutathione levels [10].

On the other side, it has to be recognised that several inflammatory molecules of natural origin exist: fructose, purines, and alkaloids should be always taken into account when speaking of toxic effects caused by the ingestion of certain vegetable products. Fresh fruits containing ≥8% of fructose might cause accelerated oxidative stress conditions in the human body. Moreover, food commodities containing purines ≤50 mg/100 g of purines are arbitrarily considered for safety concerns, while the general recommendation of consuming foods unable to determine ≤300 mg of

uric acid is valid. The problem of inflammatory alkaloids (in contrast with anti-inflammatory glycoalkaloids) is quite complex, also because blanching and juicing processes in the food industry can increase their levels in conjunction with thermal augments. Consequently, some brief indication concerning analytical detection of these chemical classes should be given in this chapter on the basis of categories: OP pesticides; ripening agents (CaC_2); fructose; purines; and alkaloids.

4.2 Natural Inflammatory Molecules Today. Conclusions and Perspectives

4.2.1 The Analytical Determination of Pesticides

In general, pesticides, including OP types, are researched and quantitatively determined in fruits and vegetables by means of gas chromatographic methods after organic extraction and concentration. At present, the most useful techniques involve gas chromatography/tandem mass spectrometry (GC-MS/MS). Actually, selective detectors such as electron capture, flame photometric, and nitrogen phosphorus detectors could be used. However, these systems can provide good sensitivity only, while specificity is generally poor. Consequently, mass spectrometry (MS) is preferred in conjunction with gas chromatography (GC). Also, tandem mass spectrometry (MS/MS) is reported to be reliable enough, especially when detectable amounts are low and original matrices are 'dirty' enough [11–22]. A good alternative method can be headspace solid-phase microextraction (HS-SPME) because it joins four analytical steps—sampling, extraction, concentration, and introduction of samples—into a single step. In addition, solvents are not used. On the other side, and in spite of rapid results and high accuracy, certain problems concerning the stability of analysed fibres and reliable sensitivity have been reported so far [23–27].

4.2.2 The Analytical Determination of Ripening Agents. The Detection of Calcium Carbide

With reference to ripening agents in fruits such as CaC_2, the following techniques—colorimetry (including an innovative gold nanoparticle-based system) method; Near-Infrared (NIR) spectroscopy; X-ray fluorescence spectrometer; elemental composition analysis (EDX) with scanning electron microscope (SEM)—have been often reported after pretreatment of fruits, washing, soaking, and acid digestion, when speaking of elemental analyses [28–30]. There are a little number of available literature concerning CaC_2 determination in fruits (calcium carbide is banned in many countries, including India). Consequently, the extension of ripening is often evaluated by means of indirect methods or the analytical determination of ripening products

in fruits. As a single example, an interesting headspace solid-phase microextraction GC–MS method has been proposed in sapota fruits with the aim of determining trithiolane residues as the result of the presence of divinyl sulphide impurities in CaC_2 [31].

4.2.3 The Analytical Determination of Fructose

With reference to fructose in fruits, a number of different techniques are available at present (fructose is often determined with other sugars at the same time). In general, the following methods have been reported in this specific ambit (and the below-mentioned list may not be fully exhaustive) [32–39]:

(a)   High-performance liquid chromatography (HPLC) systems, with joint detectors such as evaporative light scattering detection (ELSD), charged aerosol detection (CAD), and refractive index detection (RID)
(b)   NIR spectroscopy and subsequent partial least-squares (PLS) elaboration of first derivative data; Fourier Transform (FT)-NIR
(c)   Ion chromatography-pulsed amperometric detection
(d)   Refractometry, automated colorimetry, and enzymatic colorimetric systems
(e)   Amperometry and chronoamperometry.

4.2.4 The Analytical Determination of Purines

With concern to the analytical detection of purines in vegetable products, the most useful choice, bases on the inherent aqueous solubility of these compounds, appears to be HPLC with appropriated confirmation systems such as photodiode array (PDA) detectors. The use of liquid chromatography (LC)-MS has been also reported when speaking of vegetable tissues as analytical matrices. Anyway, HPLC is reported to be the first choice when speaking of reliability and sensibility [40–42].

4.2.5 The Analytical Determination of Alkaloids

The problem with alkaloids is that this term means a large group of biomolecules, also including glycoalkaloids (well-known for their anti-inflammatory properties). These compounds can have different chemical/biochemical/biological activity, depending on their biosynthetic pathways. Consequently, it could be hard to define a prominent method of analysis for 'alkaloids' in foods, especially considering the vegetable matrix. Anyway, the first steps in the analytical determination are (targeted molecules for this example are indole-, isoquinoline-, and tropane-derived alkaloids) [43]:

(a) Extraction and purification. Generally, this step is carried out by means of the following techniques: hydrodistillation; microwave-assisted extraction; pressurised liquid extraction; soxhlet extraction; ultrasound-assisted extraction; supercritical fluid extraction. The separation/extraction from the vegetable matrix may be also carried out 'traditionally', by means of digestion, infusion, decoction, and boiling under reflux. Vegetable samples have to be finely grounded, mechanically or by means of adequate sonicators, or demolished by means of non-ionic detergents [44–49].

(b) After the concentration of extracts and adequate storage (otherwise, alkaloids could be damaged), all types of chromatographic methods can be used for separation and subsequent detection [50]. It has to be remembered that GC is suitable for very volatile compounds. On the other hand, HPLC seems—and it is reported to be—the best technique in this specific ambit [51–52]. With relation to the characterisation of functional groups, different techniques—Fourier-transform infrared spectroscopy (FT-IR); MS; nuclear magnetic resonance—are available at present, and are widely used with excellent performances [53–55].

References

1. Hu F (2014) Foods that fight inflammation. Harvard Medical School, Harvard Health Publishing, Harvard. Updated 29 August 2020. https://www.health.harvard.edu/staying-healthy/foods-that-fight-inflammation. Accessed 23 June 2021
2. Lawler M (2020) A comprehensive guide to an anti-inflammatory diet. www.everydayhealth.com. https://www.everydayhealth.com/diet-nutrition/diet/anti-inflammatory-diet-benefits-food-list-tips/. Accessed 23 June 2021
3. Dryden M (2017) Reactive oxygen therapy; a novel therapy in soft tissue infection. Curr Opin Infect Dis 30(2):143–149. https://doi.org/10.1097/QCO.0000000000000350
4. Obón-Santacana M, Romaguera D, Gracia-Lavedan E, Molinuevo A, Molina-Montes E, Shivappa N, Hebert JR, Tardón A, Castaño-Vinyals G, Moratalla F, Guinó E, Marcos-Gragera R, Azpiri M, Gil L, Olmedo-Requena R, Lozano-Lorca M, Alguacil J, Fernández-Villa T, Martín V, Molina AJ, Ederra M, Moreno-Iribas C, Perez B, Aragonés N, Castello A, Huerta JM, Dierssen-Sotos T, Gómez-Acebo I, Molina-Barceló A, Pollán M, Kogevinas M, Moreno V, Amiano P (2019) Dietary inflammatory index, dietary non-enzymatic antioxidant capacity, and colorectal and breast cancer risk (MCC-Spain Study). Nutr 11(6):1406. https://doi.org/10.3390/nu11061406
5. Mayo Clinic Staff (2021) The truth behind the most popular diet trends of the moment. Mayo Clinic, Rochester. https://www.mayoclinic.org/healthy-lifestyle/weight-loss/in-depth/the-truth-behind-the-most-popular-diet-trends-of-the-moment/art-20390062. Accessed 23rd June 2021
6. Khanna S, Jaiswal KS, Gupta B (2017) Managing rheumatoid arthritis with dietary interventions. Front Nutr 4:52. https://doi.org/10.3389/fnut.2017.00052
7. Anonymous (2011) Dirt poor: have fruits and vegetables become less nutritious? www.scientificamerican.com. https://www.scientificamerican.com/article/soil-depletion-and-nutrition-loss/. Accessed 23 June 2021

8. De I, Kour A, Wani H, Sharma P, Panda JJ, Singh M (2020) Exposure of calcium carbide induces apoptosis in mammalian fibroblast L929 cells. Toxicol Mech Methods 31(3):1–11. https://doi.org/10.1080/15376516.2020.1849484
9. de Sousa BM, de Sousa MG, de Castro E, Sousa JM, Peron AP (2016) Cytotoxic and genotoxic potential of powdered juices. Food Sci Technol 36(1):49–55. https://doi.org/10.1590/1678-457x.0006
10. Owen JB, Butterfield DA (2010) Measurement of oxidized/reduced glutathione ratio. In: Totowa NJ (ed) Protein misfolding and cellular stress in disease and aging. In: Bross P, Gregersen N (eds) Protein misfolding and cellular stress in disease and aging. Methods Mol Biol Methods Protoc, vol 648, pp 269–277. https://doi.org/10.1007/978-1-60761-756-3_18
11. De Paoli M, Taccheo Barbina M, Damiano V, Fabbro D, Bruno R (1997) Simplified determination of combined residues of prochloraz and its metabolites in vegetable, fruit and wheat samples by gas chromatography. J Chromatogr A 765(1):127–131. https://doi.org/10.1016/S0021-9673(96)00995-8
12. Andersson A, Ohlin B (1986) A capillary gas chromatographic multiresidue method for determination of pesticides in fruits and vegetables. Vaar Foeda Suppl 38:79–109
13. Luke MA, Masumoto HT, Cairns T, Hundley HK (1988) Levels and incidences of pesticide residues in various foods and animal feeds analyzed by the Luke multiresidue methodology for fiscal years 1982–1986. J AOAC 71(2):415–433
14. McMahon B, Hardin N (1994) Pesticide analytical manual 1, 3rd edn. U.S. Food and Drug Administration, Washington, DC
15. Coscollá R, Gamón M (1997) The spanish pesticide residue monitoring programme: design and results. Pestic Sci 50(2):155–159. https://doi.org/10.1002/(SICI)1096-9063(199706)50:2%3c111::AID-PS573%3e3.0
16. Plomley JB, Koester CJ, March RE (1994) Determination of N-nitrosodimethylamine in complex environmental matrixes by quadrupole ion storage tandem mass spectrometry enhanced by unidirectional ion ejection. Anal Chem 66(24):4437–4443. https://doi.org/10.1021/ac00096a008
17. Hayward DG, Hooper K, Andrzejewski D (1999) Tandem-in-time mass spectrometry method for the sub-parts-per-trillion determination of 2, 3, 7, 8-chlorine-substituted dibenzo-p-dioxins and-furans in high-fat foods. Anal Chem 71(1):212–220. https://doi.org/10.1021/ac980282+
18. Contreras M, Mocholi F (2003) Analysis of non-volatile pesticide residue invegetables by positive/negative ESI LC/MS/MS. LC/MS Application Note 11. http://varianinc.com/, https://www.agilent.com/cs/library/applications/lcms11.pdf. Accessed 26 June 2021
19. Schachterle S, Feigel C (1996) Pesticide residue analysis in fresh produce by gas chromatography-tandem mass spectrometry. J Chromatogr A 754(1–2):411–422. https://doi.org/10.1016/S0021-9673(96)00492-X
20. Sheridan RS, Meola JR (1999) Analysis of pesticide residues in fruits, vegetables, and milk by gas chromatography/tandem mass spectrometry. J AOAC Int 82(4):982–990. https://doi.org/10.1093/jaoac/82.4.982
21. Bempah CK, Donkor A, Yeboah PO, Dubey B, Osei-Fosu P (2011) A preliminary assessment of consumer's exposure to organochlorine pesticides in fruits and vegetables and the potential health risk in Accra Metropolis. Ghana. Food Chem 128(4):1058–1065. https://doi.org/10.1016/j.foodchem.2011.04.013
22. Gamón M, Lleó C, Ten A, Mocholí F (2001) Multiresidue determination of pesticides in fruit and vegetables by gas chromatography/tandem mass spectrometry. J AOAC Int 84(4):1209–1216. https://doi.org/10.1093/jaoac/84.4.1209
23. Volante M, Pontello M, Cattaneo M, Calzoni L (2000) Application of solid phase micro-extraction (SPME) to the analysis of pesticides residues in vegetables. Pest Manag Sci 56(7):618–636. https://doi.org/10.1002/1526-4998(200007)56:7%3C618::AID-PS178%3E3.0.CO;2-E
24. Pihlstrom T, Osterdahl B (1999) Analysis of pesticide residues in fruit and vege-tables after cleanup with solid-phase extraction using ENV + (polystyrene–divinylbenzene) cartridges. J Agr Food Chem 47(7):2549–2552. https://doi.org/10.1021/jf981393c

25. Lambropoulou D, Albanis T (2003) Headspace solid-phase microextraction in combination with gas chromatography-mass spectrometry for the rapid screen-ing of organophosphorus insecticide residues in strawberries and cherries. J Chromatogr A 993(1–2):197–203. https://doi.org/10.1016/S0021-9673(03)00397-2

26. Fytianos K, Raikos N, Theodoridis G, Velinova Z, Tsoukali H (2006) Solid phase microextraction applied to the analysis of organophosphorus insecticides in fruits. Chemosph 65(11):2090–2095. https://doi.org/10.1016/j.chemosphere.2006.06.046

27. Lakade AJ, Sundar K, Shetty PH (2018) Gold nanoparticle-based method for detection of calcium carbide in artificially ripened mangoes (Magnifera indica). Food Addit Contam A 35(6):1078–1084. https://doi.org/10.1080/19440049.2018.1449969

28. Islam MN, Imtiaz MY, Alam SS, Nowshad F, Shadman SA, Khan MS (2018) Artificial ripening on banana (Musa Spp.) samples: analyzing ripening agents and change in nutritional parameters. Cogent Food Agric 4, 1:1477232. https://doi.org/10.1080/23311932.2018.1477232

29. Palpandian P, Shanmugam H, Rani EA, Prabu GTV (2019) Determination of fruit quality of calcium carbide induced ripening in mango (Mangifera indica L. cv. Alphonso) by physiological, biochemical, bio-enzymatic and elemental composition analysis (EDX). Indian J Biochem Biophys 56(3):205–213

30. Mahmood T, Saeed I, Anwer H, Mahmood I, Zubair A (2013) Comparative study to evaluate the effect of calcium carbide (CaC₂) as an artificial ripening agent on shelf life, physio-chemical properties, iron containment and quality of prunus persica l. Batsch. Eur Acad Res 1(5):685–700

31. Vemula M, Shaikh AS, Chilakala S, Tallapally M, Upadhyayula V (2020) Identification of calcium carbide-ripened sapota (Achras sapota) fruit by headspace SPME-GC-MS. Food Addit Contam A 37(10):1601–1609. https://doi.org/10.1080/19440049.2020.1794055

32. Rodriguez-Saona LE, Fry FS, McLaughlin MA, Calvey EM (2001) Rapid analysis of sugars in fruit juices by FT-NIR spectroscopy. Carbohydr Res 336(1):63–74. https://doi.org/10.1016/S0008-6215(01)00244-0

33. Beilmann B, Langguth P, Häusler H, Grass P (2006) High-performance liquid chromatography of lactose with evaporative light scattering detection, applied to determine fine particle dose of carrier in dry powder inhalation prod-ucts. J Chromatogr A 1107(1):204–207. https://doi.org/10.1016/j.chroma.2005.12.083

34. Ma C, Sun Z, Chen C, Zhang L, Zhu S (2014) Simultaneous separation and determination of fructose, sorbitol, glucose and sucrose in fruits by HPLC–ELSD. Food Chem 145:784–788. https://doi.org/10.1016/j.foodchem.2013.08.135

35. Romero-Rodriguez MA, Vazquez-Oderiz ML, Lopez-Hernandez J, Simal-Lozano J (1992) Physical and analytical characteristics of the kiwano. J Food Comp Anal 5(4):319–322. https://doi.org/10.1016/0889-1575(92)90065-R

36. Aksorn J, Teepoo S (2020) Development of the simultaneous colorimetric enzymatic detection of sucrose, fructose and glucose using a microfluidic paper-based analytical device. Talanta 207:120302. https://doi.org/10.1016/j.talanta.2019.120302

37. Wang B, Wang X, Bei J, Xu L, Zhang X, Xu Z (2021) Devel-opment and validation of an analytical method for the quantification of arabinose, galactose, glucose, sucrose, fructose, and maltose in fruits, vegetables, and their products. Food Anal Methods 14:1227–1238. https://doi.org/10.1007/s12161-021-01964-y

38. Calull M, Marce R, Borrull F (1992) Determination of carboxylic acids, sugars, glycerol and ethanol in wine and grape must by ion-exchange high performance liquid chromatography with refractive index detection. J Chromatogr A 590(2):215–222. https://doi.org/10.1016/0021-9673(92)85384-6

39. Rambla FJ, Garrigues S, De La Guardia M (1997) PLS-NIR determination of total sugar, glucose, fructose and sucrose in aqueous solutions of fruit juices. Anal Chimica Acta 344(1–2):41–53. https://doi.org/10.1016/S0003-2670(97)00032-9

40. Lontoc AV, Ladines EO, Jalandon SA (1993) Purine content of some Philippine foods. Kimika 9:15–22

41. Yamaoka N, Kaneko K, Kudo Y, Aoki M, Yasuda M, Mawatari K, MawatariK NK, Yamada Y, Yamamoto T (2010) Analysis of purine in purine-rich cauliflower. Nucleosides Nucleotides Nucl Acids 29(4–6):518–521. https://doi.org/10.1080/15257771003741372

42. Rong S, Zou L, Wang Z, Pan H, Yang Y (2012) Purine in common plant food in China. J Hygiene Res 41(1):92–95

43. Dey P, Kundu A, Kumar A, Gupta M, Lee BM, Bhakta T, Dash S, Kim HS (2020) Analysis of alkaloids (indole alkaloids, isoquinoline alkaloids, tropane alkaloids). In: Nabavi SM, Saeedi M, Nabavi SF, Sanches Silva A (eds) Recent advances in natural products analysis, pp 505–567. Elsevier, Amsterdam. https://doi.org/10.1016/B978-0-12-816455-6.00015-9

44. Bhargavi G, Rao PN, Renganathan S (2018) Review on the Extraction Methods of Crude oil from all Generation Biofuels in last few Decades. In: IOP conference series: materials science and engineering, vol 330, no 1, p 012024. IOP Publishing, Bristol. https://doi.org/10.1088/1757-899X/330/1/012024

45. Kirakosyan A, Noon KR, McKenzie M, Cseke LJ, Kaufman PB (2016) Isolation and Purification of Metabolites. In: Cseke LJ, Ara K, Kaufman PB, Westfall MV (eds) Handbook of Molecular and Cellular Methods in Biology and Medicine, pp 386–399. https://doi.org/10.1201/b11351-25

46. Huie CW (2002) A review of modern sample-preparation techniques for the extraction and analysis of medicinal plants. Anal Bioanal Chem 373(1):23–30. https://doi.org/10.1007/s00216-002-1265-3

47. Fabricant DS, Farnsworth NR (2001) The value of plants used in traditional medicine for drug discovery. Environ Health Perspect 109, Supplement 1:69–75.https://doi.org/10.1289/ehp.01109s169

48. Özek G, Demirci F, Özek T, Tabanca N, Wedge D, Khan S, Başer KHC, Duran A, Hamzaoglu E (2010) Gas chromatographic-mass spectrometric analysis of volatiles obtained by four different techniques from Salvia rosifolia Sm., and evaluation for biological activity. J Chromatogr A 1217(5):741–748. https://doi.org/10.1016/j.chroma.2009.11.086

49. Farhat A, Ginies C, Romdhane M, Chemat F (2009) Eco-friendly and cleaner process for isolation of essential oil using microwave energy: experimental and theoretical study. J Chromatogr A 1216(26):5077–5085. https://doi.org/10.1016/j.chroma.2009.04.084

50. Coskun O (2016) Separation techniques: chromatography. North Clin Istanb 3(2):156–160. https://doi.org/10.14744/nci.2016.32757

51. Swartz M (2010) HPLC detectors: a brief review. J Liq Chromatogr Relat Technol 33(9–12):1130–1150. https://doi.org/10.1080/10826076.2010.484356

52. Thammana M (2016) A review on high performance liquid chromatography (HPLC). Res Rev J Pharm Anal 5(2):22–28

53. Schripsema J (2010) Application of NMR in plant metabolomics: techniques, problems and prospects. Phytochem Anal Int J Plant Chem Biochem Tech 21(1):14–21. https://doi.org/10.1002/pca.1185

54. Rammohan B, Samit K, Chinmoy D, Arup S, Amit K, Ratul S, Sanmoy K, Dipan A, Tuhinadri S (2016) Human cytochrome P450 enzyme modulation by Gymnema sylvestre: a predictive safety evaluation by LC-MS/MS. Pharmacogn Mag 12, Suppl 4:S389–S394. https://doi.org/10.4103/0973-1296.191441

55. Hazra K, Roy R, Sen S, Laskar S (2007) Isolation of antibacterial pentahydroxy flavones from the seeds of Mimusops elengi Linn. Afr J Biotechnol 6(12):1446–1449

Chapter 5
Selection of Fruits and Vegetables with Public Health Viewpoint: Discussion

Abstract Proteins and lipids are molecules that might be considered as soft targets for oxidative attack, whose modification in the body can increase the risk of inflammation due to mutagenesis. Effects of food processing additives with oxidant attitude are known. Consequently, oxidation during food processing may be a problem, especially when speaking of high-energy foods derived from cereals, with notable amounts of oxidation-sensible lipids and protein chains. The public health perspective with relation to some natural molecules which might lead to excessive oxidation reactions in the human body is extremely important. Consequently, some discussion should be dedicated to unsaturated fatty acids and some of the known polypeptides in cereal-based high-energy foods. Processed high-energy bars and other snacks derived from cereals are really 'energetic' if compared with whole cereal grains, and the palatability is directly proportional to the amount of sugars (sweetness) and fats/oils. However, other problems could occur when speaking of high-energy foods. Ultraprocessed foods are completely dissimilar from original foods, and these products contain a notable amount of oxidation-sensible molecules: lipids and proteins. Consequently, processing and non-processing factors could have some important public safety consequences. The aim of this chapter is to give some perspective to this ambit.

Keywords High-energy food · Maillard reaction · Oxidation · Polyunsaturated fatty acid · Reactive nitrogen species · Reactive oxygen species · Ready-to-eat

Abbreviations

H_2O_2	Hydrogen peroxide
$\cdot OH^-$	Hydroxyl radical
NO	Nitric oxide
R-OO·	Peroxyl radical
PUFA	Polyunsaturated fatty acid
RNS	Reactive nitrogen species
ROS	Reactive oxygen species

© The Author(s), under exclusive license to Springer Nature Switzerland AG 2022   49
R. K. Sharma et al., *Natural Inflammatory Molecules in Fruits and Vegetables*,
Chemistry of Foods, https://doi.org/10.1007/978-3-030-88473-4_5

RTE   Ready-to-eat
$\cdot O_2^-$ Singlet oxygen

5.1   An Introduction to High-Energy Foods

Excessive reactive oxygen species (ROS) can lead to oxidative stress and consequently permanent inflammation in the human body, mediating chronic diseases like
cancer, diabetes, cardiovascular, neurological, and pulmonary diseases (Chap. 1). In
addition, the mitochondrial respiratory chain can produce under hypoxic conditions
nitric oxide (NO). This molecule is able to generate reactive nitrogen species (RNS),
with the consequent possibility of lipid peroxidation.

Proteins and lipids are particular molecules that might be considered as soft targets
for oxidative attack, whose modification in the body can increase the risk of inflammation due to mutagenesis. Effects of processing additives with oxidant attitude, such
as benzoyl peroxide in bread or chlorine dioxide in flour, are known. Consequently,
oxidation during food processing may be a problem, especially when speaking of
high-energy foods (derived from cereals) with notable amounts of oxidation-sensible
lipids and protein chains. The public health perspective with relation to some natural
molecules, found in fruits and vegetables and their derivatives (also fats and oils),
which might lead to excessive oxidation reactions in the human body, is extremely
important, at present. Consequently, some discussion should be dedicated to unsaturated fatty acids and some of the known polypeptides in cereal-based high-energy
foods because of their liability to NO production.

On the one side, cereals can be used as whole raw materials (whole grains) because
of their low cost (if compared with processed foods from cereals). Economic cheapness is a good thing, but normal consumers seem to be more interested in high-
palatable foods such as energy snacks and so on. The problem—actually, only one of
the possible problems—is that processed foods and beverages 'value' a huge number
of calories because of the disproportionate presence of sugars and fats of vegetable
or animal origin [1]. For these reasons, processed high-energy bars and other snacks
derived from cereals are really 'energetic' if compared with whole cereal grains, also
in form of ready-to-eat (RTE) breakfast cereals [2]. In addition, palatability is directly
proportional to the amount of sugars (sweetness) and fats/oils (these molecules may
have a synergetic effect if in conjunction with added sugars when speaking of sweet
foods). Moreover, the caloric intake is enormously augmented when speaking of
high-energy food consumption, because a full-satiation effect is reachable in this way
only after a certain (and excessive) consumption, also depending on non-nutritional
factors [3–6].

On the other side, unprocessed cereals and other vegetable foods without prior
processing are generally able to give full-satiation effects with a low consumption
ratio if compared with high-energy snacks obtained from the same raw materials. This

behaviour has been reported with concern to whole grains, raw almonds, satiereal, and minimally processed foods [2–11].

Anyway, the positive correlation between high-energy food and metabolic disorders such as obesity has been defined and reported with notable reliability [12]. However, the public health perspective should consider not only this type of correlation. Other problems could occur when speaking of high-energy foods, especially considering that these products are the result of more than one single processing activity. In fact, processed foods could be classified as follows (Fig. 5.1) [13]:

(a) Minimally processed foods,
(b) Extracted ingredients,
(c) Ultraprocessed foods.

In general, minimally processed foods are not a big issue because of a slight difference—from the nutritional viewpoint—with whole foods (original raw materials). Processing options aim at the preservation of nutritional features, while the properties such as cheapness, safety, palatability should be enhanced. Anyway, processing options should not alter the original composition because the addition of supplemental ingredients/additives is not desired [13].

On the other side, certain oils, fats, and other nutritional groups (sugars, proteins…) can be extracted from raw materials of vegetable origin, and this group concerns substantially ingredients (example; palm oil, coconut oil, …). Obviously,

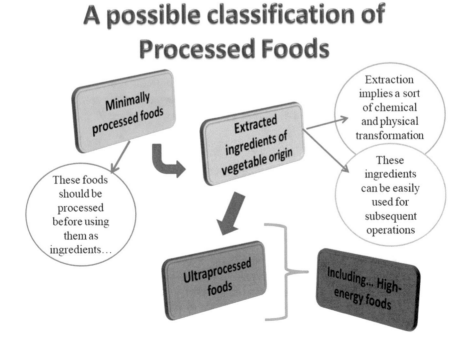

Fig. 5.1 A possible classification of processed foods in the modern food industry

being they processed ingredients, these commodities can be used in the production of ultraprocessed foods because of two features [13]:

(1)   Extraction implies a sort of chemical and physical transformation
(2)   Because of chemical/physical transformation, these ingredients can be easily used for subsequent operations, while original raw materials could not be so good.

Finally, ultraprocessed foods—including high-energy products from cereals—are completely dissimilar from original foods, unless marketing choices suggest that a sort of similarity is required by the normal consumer. Anyway, these products should be considered as a concentrated source of high-energy chemical compounds (in terms of supplied energy per volumetric amount of served food), while sugars, proteins, and lipids have a notable abundance. The introduction of 'light' ultraprocessed foods with lower sugar contents, lower trans fat quantities, etc. cannot be considered as a good alternative, because the comparison is made between two ultraprocessed foods at the same time [13].

Another problem with high-energy foods is that these products contain a notable amount of oxidation-sensible molecules: lipids and proteins. Consequently, processing and non-processing factors (storage in humid areas, high temperature, simple air oxidation in unprotected conditions, etc.) could have some important public safety consequences in this ambit because of the processed nature of selected nutrients on the one side, and of the notable amount in these products.

5.2   Lipids and Proteins as Oxidation Targets

Protein and lipid molecules of vegetable origin are high-energy factors from the nutritional viewpoint. Also, should these compounds be excessively consumed, they could also be good targets for oxidative attack with one or more of the following diseases (Chap. 1):

(a)   Atherosclerosis,
(b)   Cardiovascular diseases, including hypertension,
(c)   Diabetes,
(d)   Dysfunction of the respiratory system, including asthma [14],
(e)   Neurological disorders,
(f)   Tumours.

By the biochemical angle, these targets are attacked with the aim of producing free radicals—ROS and RNS—useful against bacterial menaces (Gram-positive and Gram-negative organisms) and viruses [15]. However, a phenomenon such as oxidative stress is the result of the general abundance of ROS and RNS if these species

—produced in presence of abundant lipids and proteins—are not properly and sufficiently balanced by enzymatic antioxidants (examples: catalase, glutathione peroxidase, superoxide dismutase...) [16, 17]. In these conditions, infection could still occur.

In detail, and with exclusive relation to high-energy foods from vegetable sources (especially cereals), the list of possible exceptions to the 'general rule' which considers vegetables and fruits as safe/healthy foods, the category of vegan products could be considered because of the notable amount of starch and vegetable (sunflower, palm, and coconut oils). Consequently, some attention should be done when speaking of the most probable targets of oxidation—lipids and proteins—of vegetable origin, when incorporated in high-energy foods. Two examples of ROS oxidation attack can be discussed.

Generated ROS and RNS are generally considered [18]:

(a) The singlet oxygen ($\cdot O_2^-$),
(b) The hydroxyl radical ($\cdot OH^-$),
(c) Hydrogen peroxide (H_2O_2),
(d) NO.

With reference to the first three ROS, one of the preferred ROS targets in high-energy foods, depending on favourable conditions, are polyunsaturated fatty acids (PUFA): this chain of reactions—lipid peroxidation—produces an alkyl radical and the correspondent hydroperoxidised lipid after the attack of a hydroxyl radical [18, 19]. Subsequently, the produced radical reacts with oxygen creating another R-OO· which should be able to attack another polyunsaturated fatty acid with the production of the hydroperoxide and a subsequent alkyl radical. Propagation continues and it is also enhanced when the number of fatty unsaturations grows. For these reasons, all oils used in high-energy foods containing these unsaturated fatty acids—such as arachidonic acid—are 'under oxidation risk', and the same thing is true for the whole food. Atherosclerosis is generally correlated with lipid peroxidation. What could be the right solution? Unfortunately, high-energy food containing vegetable oils are all subjected to lipid peroxidation; consequently, different strategies may be used—low processing temperatures; slightly augmented water (with the aim of increasing activity water); elimination of chlorophylls and riboflavin (as photosensitisers)—but it could be difficult. Paradoxically, the increase of saturated fats could help, but it is questionable for public health purposes. Another strategy is linked to edible coating solutions for surface protection against oxidation. Anyway, all used oils and fats of vegetable origin are surely 'under oxidation risk'. Interestingly, the role of proteins in lipid peroxidation is not clear because some researches imply that proteins could prevent partially oxidative damages for fat molecules [20].

In addition, lipid peroxidation and similar oxidative chain reactions can produce toxic aldehydes such as malonaldehyde. This molecule is particularly able to attack the amino group of proteins, with particular relation to the possible covalent bond formed between two lysine residues from two different proteins. Actually, this type of reaction can occur when two amino groups of the same protein are available. Anyway, the result is the production of Schiff bases [21–23], and the possibility of heterocycles can occur. On these bases, it can be inferred that the abundance of proteins containing relevant amounts of lysine residues could be 'under oxidation risk'. It has to be considered also that proteins can undergo three different oxidation attacks: oxidation of amino acid residues, breaking of peptide bonds, and covalent links between different proteins (with consequent agglomeration) [18]. Anyway, the propagation occurs when alkyl radicals can continue to react with oxygen, while the low abundance of oxygen forces proteins to agglomerate. With relation to high-energy food from vegetable sources, these risks have to be always taken into account, even if processing options—and water removal—could easily suggest more visible phenomena: Maillard reaction products and caramelisation above all [21–23]. On the other side, oxidation phenomena on proteins could be less visible and have sure effects on human health. Consequently, more research is needed in this ambit yet [24–26]. At present, a notable part of studies on protein oxidation concerns research on animal (meat substrates), processed meats, and fish foods [27]. However, several studies are extremely promising on vegetable proteins, showing that one of the most interesting results of ROS attacks is linked with conformational and functional features of oxidised proteins [28].

References

1. Popkin BM (2011) Agricultural policies, food and public health. EMBO Rep 12(1):11–18. https://doi.org/10.1038/embor.2010.200
2. Miller HE, Rigelhof F, Marquart L, Prakash A, Kanter M (2000) Antioxidant content of whole grain breakfast cereals, fruits and vegetables. J Am Coll Nutr 19(Suppl 3):312S–319S. https://doi.org/10.1080/07315724.2000.10718966
3. Brogden N, Almiron-Roig E (2010) Food liking, familiarity and expected satiation selectively influence portion size estimation of snacks and caloric beverages in men. Appetite 55(3):551–555. https://doi.org/10.1016/j.appet.2010.09.003
4. Dougkas A, Minihane AM, Givens DI, Reynolds CK, Yaqoob P (2012) Differential effects of dairy snacks on appetite, but not overall energy intake. Br J Nutr 108(12):2274–2285. https://doi.org/10.1017/S0007114512000323
5. Martin AA, Hamill LR, Davies S, Rogers PJ, Brunstrom JM (2015) Energy-dense snacks can have the same expected satiation as sugar-containing beverages. Appetite 95:81–88. https://doi.org/10.1016/j.appet.2015.06.007
6. Lange C, Schwartz C, Hachefa C, Cornil Y, Nicklaus S, Chandon P (2020) Portion size selection in children: effect of sensory imagery for snacks varying in energy density. Appetite 150:104656. https://doi.org/10.1016/j.appet.2020.104656
7. Hull S, Re R, Chambers L, Echaniz A, Wickham MS (2015) A mid-morning snack of almonds generates satiety and appropriate adjustment of subsequent food intake in healthy women. Eur J Nutr 54(5):803–810. https://doi.org/10.1007/s00394-014-0759-z
8. Gout B, Bourges C, Paineau-Dubreuil S (2010) Satiereal, a Crocus sativus L extract, reduces snacking and increases satiety in a randomized placebo-controlled study of mildly overweight, healthy women. Nutr Res 30(5):305–313. https://doi.org/10.1016/j.nutres.2010.04.008
9. Green SM, Blundell JE (1996) Subjective and objective indices of the satiating effect of foods. Can people predict how filling a food will be? Eur J Clin Nutr 50(12):798–806
10. Fardet A (2016) Minimally processed foods are more satiating and less hyperglycemic than ultra-processed foods: a preliminary study with 98 ready-to-eat foods. Food Funct 7(5):2338–2346. https://doi.org/10.1039/C6FO00107F

11. Hollingworth S, Dalton M, Blundell JE, Finlayson G (2019) Evaluation of the influence of raw almonds on appetite control: satiation, satiety, hedonics and consumer perceptions. Nutrients 11(9):2030. https://doi.org/10.3390/nu11092030

12. Darmon N, Briend A, Drewnowski A (2004) Energy-dense diets are associated with lower diet costs: a community study of French adults. Pub Health Nutr 7(1):21–27. https://doi.org/10.1079/PHN2003512

13. Monteiro CA (2009) Nutrition and health. The issue is not food, nor nutrients, so much as processing. Pub Health Nutr 12(5):729–731. https://doi.org/10.1017/S1368980009005291

14. Birben E, Sahiner UM, Sackesen C, Erzurum S, Kalayci O (2012) Oxidative stress and antioxidant defense. J World Allergy Organ 5(1):9–19. https://doi.org/10.1097/WOX.0b013e318243 9613

15. Dryden M (2017) Reactive oxygen therapy; a novel therapy in soft tissue infection. Curr Opin Infect Dis 30(2):143–149. https://doi.org/10.1097/QCO.0000000000000350

16. Deponte M (2013) Glutathione catalysis and the reaction mechanism of glutathione-dependent enzymes. Biochim Biophys Acta 1830(5):3217–3266. https://doi.org/10.1016/j.bagen.2012. 09.018

17. Pizzino G, Irrera N, Cucinotta M, Pallio G, Mannino F, Arcoraci V, Squadrito F, Altavilla D, Bitto A (2017) Oxidative stress—harms and benefits for human health. Oxid Med Cell Longev 2017:8416763. https://doi.org/10.1155/2017/8416763

18. Juan CA, Pérez de la Lastra JM, Plou FJ, Pérez-Lebeña E (2021) The chemistry of Reactive Oxygen Species (ROS) revisited: outlining their role in biological macromolecules (DNA, lipids and proteins) and induced pathologies. Int J Mol Sci 22(9):4642. https://doi.org/10.3390/ijm s22094642

19. Yadav DK, Kumar S, Choi EH, Chaudhary S, Kim MH (2019) Molecular dynamic simulations of oxidized skin lipid bilayer and permeability of reactive oxygen species. Sci Rep 9:4496. https://doi.org/10.1038/s41598-019-40913-y

20. Barden L, Decker EA (2016) Lipid oxidation in low-moisture food: a review. Crit Rev Food Sci Nutr 56(15):2467–2482. https://doi.org/10.1080/10408398.2013.848833

21. Parisi S, Luo W (2018) Chemistry of Maillard reactions in processed foods. Springer International Publishing, Cham. https://doi.org/10.1007/978-3-319-95463-9

22. Singla RK, Dubey AK, Ameen SM, Montalto S, Parisi S (2018) Analytical methods for the assessment of Maillard reactions in foods. Springer International Publishing, Cham. https://doi.org/10.1007/978-3-319-76923-3_1

23. Parisi S, Ameen SM, Montalto S, Santangelo A (2019) Maillard reaction in foods. Springer International Publishing, Cham. https://doi.org/10.1007/978-3-030-22556-8_3

24. Hellwig M (2019) The chemistry of protein oxidation in food. Angew Chem Int Ed 58(47):16742–16763. https://doi.org/10.1002/anie.201814144

25. Lund MN, Baron CP (2010) Protein oxidation in foods and food quality. In: Skibsted LH, Risbo J, Andersen ML (eds) Chemical deterioration and physical instability of food and beverages. Woodhead Publishing Ltd., Sawston, pp 33–69. https://doi.org/10.1533/9781845699260.1.33

26. Estévez M, Luna C (2017) Dietary protein oxidation: a silent threat to human health? Crit Rev Food Sci Nutr 57(17):3781–3793. https://doi.org/10.1080/10408398.2016.1165182

27. Jongberg S, Lund MN, Skibsted LH (2017) Protein oxidation in meat and meat products. Challenges for antioxidant protection. In: Barbosa-Cánovas GV, Pastore GM, Candoğan K, Medina Meza IG, Caetano da Silva Lannes S, Buckle K, Yada RY, Rosenthal A (eds) Global food security and wellness. Springer, New York, pp 315–337. https://doi.org/10.1007/978-1-4939-6496-3_17

28. Ye L, Liao Y, Zhao M, Sun W (2013) Effect of protein oxidation on the conformational properties of peanut protein isolate. J Chem 2013(Article ID 423254). https://doi.org/10.1155/2013/423254

Printed in the United States
by Baker & Taylor Publisher Services